VOLUME SEVENTY SEVEN

ADVANCES IN
MARINE BIOLOGY

Northeast Pacific Shark Biology, Research and Conservation Part A

ADVANCES IN MARINE BIOLOGY

VOLUME SEVENTY SEVEN

ADVANCES IN MARINE BIOLOGY

Northeast Pacific Shark Biology, Research and Conservation Part A

Edited by

SHAWN E. LARSON
Seattle Aquarium
Seattle, WA
United States

DAYV LOWRY
Marine Fish Science Unit
Washington Department of Fish and Wildlife
Olympia, WA
United States

ELSEVIER

ACADEMIC PRESS
An imprint of Elsevier

Academic Press is an imprint of Elsevier
125 London Wall, London, EC2Y 5AS, United Kingdom
The Boulevard, Langford Lane, Kidlington, Oxford OX5 1GB, United Kingdom
525 B Street, Suite 1800, San Diego, CA 92101-4495, United States
50 Hampshire Street, 5th Floor, Cambridge, MA 02139, United States

First edition 2017

Notices
Knowledge and best practice in this field are constantly changing. As new research and experience broaden our understanding, changes in research methods, professional practices, or medical treatment may become necessary.

Practitioners and researchers must always rely on their own experience and knowledge in evaluating and using any information, methods, compounds, or experiments described herein. In using such information or methods they should be mindful of their own safety and the safety of others, including parties for whom they have a professional responsibility.

To the fullest extent of the law, neither the Publisher nor the authors, contributors, or editors, assume any liability for any injury and/or damage to persons or property as a matter of products liability, negligence or otherwise, or from any use or operation of any methods, products, instructions, or ideas contained in the material herein.

ISBN: 978-0-12-811831-3
ISSN: 0065-2881

Publisher: Zoe Kruze
Acquisition Editor: Jason Mitchell
Editorial Project Manager: Helene Kabes
Production Project Manager: Vignesh Tamilselvan
Senior Cover Designer: Christian J. Bilbow

Typeset by SPi Global, India

CONTRIBUTORS TO VOLUME 77

Jennifer S. Bigman
Earth to Ocean Research Group, Simon Fraser University, Burnaby, BC, Canada

Joseph J. Bizzarro
Moss Landing Marine Laboratories, Moss Landing, CA, United States

Aaron B. Carlisle
Hopkins Marine Station of Stanford University, Pacific Grove, CA, United States

Enric Cortés
National Marine Fisheries Service, Southeast Fisheries Science Center, Panama City Laboratory, FL, United States

Toby S. Daly-Engel
Florida Institute of Technology, Melbourne, FL, United States

David A. Ebert
Pacific Shark Research Center, Moss Landing Marine Laboratories, Moss Landing, CA, United States

Christopher M. Gburski
Resource Ecology and Fisheries Management, Alaska Fisheries Science Center, National Marine Fisheries Service, National Oceanic and Atmospheric Administration, Seattle, WA, United States

Kenneth J. Goldman
Alaska Department of Fish and Game, Homer, AK, United States

Chris J. Harvey
National Oceanic and Atmospheric Administration, Northwest Fisheries Science Center, Seattle, WA, United States

Shawn E. Larson
Seattle Aquarium, Seattle, WA, United States

Julia M. Lawson
Earth to Ocean Research Group, Simon Fraser University, Burnaby, BC, Canada

Dayv Lowry
Marine Fish Science Unit, Washington Department of Fish and Wildlife, Olympia, WA, United States

Mary E. Matta
Resource Ecology and Fisheries Management, Alaska Fisheries Science Center, National Marine Fisheries Service, National Oceanic and Atmospheric Administration, Seattle, WA, United States

Nicole M. Phillips
The University of Southern Mississippi, Hattiesburg, MS, United States

Jonathan C.P. Reum
University of Washington, School of Aquatic and Fishery Sciences, Seattle, WA, United States

Wade D. Smith
University of British Columbia, Institute for the Oceans and Fisheries, Vancouver, BC, Canada

Cindy A. Tribuzio
Auke Bay Laboratories, Alaska Fisheries Science Center, National Marine Fisheries Service, National Oceanic and Atmospheric Administration, Juneau, AK, United States

Gregory D. Williams
Pacific States Marine Fisheries Commission, Seattle, WA, United States

CONTENTS

SERIES CONTENTS FOR LAST FIFTEEN YEARS*

Volume 38, 2000.

Blaxter, J. H. S. The enhancement of marine fish stocks. pp. 1–54.

Bergström, B. I. The biology of Pandalus. pp. 55–245.

Volume 39, 2001.

Peterson, C. H. The "Exxon Valdez" oil spill in Alaska: acute indirect and chronic effects on the ecosystem. pp. 1–103.

Johnson, W. S., Stevens, M. and Watling, L. Reproduction and development of marine peracaridans. pp. 105–260.

Rodhouse, P. G., Elvidge, C. D. and Trathan, P. N. Remote sensing of the global light-fishing fleet: an analysis of interactions with oceanography, other fisheries and predators. pp. 261–303.

Volume 40, 2001.

Hemmingsen, W. and MacKenzie, K. The parasite fauna of the Atlantic cod, Gadus morhua L. pp. 1–80.

Kathiresan, K. and Bingham, B. L. Biology of mangroves and mangrove ecosystems. pp. 81–251.

Zaccone, G., Kapoor, B. G., Fasulo, S. and Ainis, L. Structural, histochemical and functional aspects of the epidermis of fishes. pp. 253–348.

Volume 41, 2001.

Whitfield, M. Interactions between phytoplankton and trace metals in the ocean. pp. 1–128.

Hamel, J.-F., Conand, C., Pawson, D. L. and Mercier, A. The sea cucumber Holothuria scabra (Holothuroidea: Echinodermata): its biology and exploitation as beche-de-Mer. pp. 129–223.

Volume 42, 2002.

Zardus, J. D. Protobranch bivalves. pp. 1–65.

Mikkelsen, P. M. Shelled opisthobranchs. pp. 67–136.

Reynolds, P. D. The Scaphopoda. pp. 137–236.

Harasewych, M. G. Pleurotomarioidean gastropods. pp. 237–294.

*The full list of contents for volumes 1–37 can be found in volume 38

Volume 43, 2002.
Rohde, K. Ecology and biogeography of marine parasites. pp. 1–86.
Ramirez Llodra, E. Fecundity and life-history strategies in marine invertebrates. pp. 87–170.
Brierley, A. S. and Thomas, D. N. Ecology of southern ocean pack ice. pp. 171–276.
Hedley, J. D. and Mumby, P. J. Biological and remote sensing perspectives of pigmentation in coral reef organisms. pp. 277–317.

Volume 44, 2003.
Hirst, A. G., Roff, J. C. and Lampitt, R. S. A synthesis of growth rates in epipelagic invertebrate zooplankton. pp. 3–142.
Boletzky, S. von. Biology of early life stages in cephalopod molluscs. pp. 143–203.
Pittman, S. J. and McAlpine, C. A. Movements of marine fish and decapod crustaceans: process, theory and application. pp. 205–294.
Cutts, C. J. Culture of harpacticoid copepods: potential as live feed for rearing marine fish. pp. 295–315.

Volume 45, 2003.
Cumulative Taxonomic and Subject Index.

Volume 46, 2003.
Gooday, A. J. Benthic foraminifera (Protista) as tools in deep-water palaeoceanography: environmental influences on faunal characteristics. pp. 1–90.
Subramoniam, T. and Gunamalai, V. Breeding biology of the intertidal sand crab, Emerita (Decapoda: Anomura). pp. 91–182.
Coles, S. L. and Brown, B. E. Coral bleaching—capacity for acclimatization and adaptation. pp. 183–223.
Dalsgaard J., St. John M., Kattner G., Müller-Navarra D. and Hagen W. Fatty acid trophic markers in the pelagic marine environment. pp. 225–340.

Volume 47, 2004.
Southward, A. J., Langmead, O., Hardman-Mountford, N. J., Aiken, J., Boalch, G. T., Dando, P. R., Genner, M. J., Joint, I., Kendall, M. A., Halliday, N. C., Harris, R. P., Leaper, R., Mieszkowska, N., Pingree, R. D., Richardson, A. J., Sims, D.W., Smith, T., Walne, A. W. and Hawkins, S. J. Long-term oceanographic and ecological research in the western English Channel. pp. 1–105.

Recommended Research Priorities in the Mediterranean Sea. pp. 109–160.

Volume 72, 2015.

Thomas A. Jefferson and Barbara E. Curry. Humpback Dolphins: A Brief Introduction to the Genus *Sousa*. pp. 1–16.

Sarah Piwetz, David Lundquist, and Bernd Würsig. Humpback Dolphin (Genus *Sousa*) Behavioural Responses to Human Activities. pp. 17–46.

Tim Collins. Re-assessment of the Conservation Status of the Atlantic Humpback Dolphin, *Sousa teuszii* (Kükenthal, 1892) Using the IUCN Red List Criteria. pp. 47–78.

Caroline R. Weir and Tim Collins. A Review of the Geographical Distribution and Habitat of the Atlantic Humpback Dolphin (*Sousa teuszii*). pp. 79–118.

Gill T Braulik, Ken Findlay, Salvatore Cerchio, and Robert Baldwin. Assessment of the Conservation Status of the Indian Ocean Humpback Dolphin (*Sousa plumbea*) Using the IUCN Red List Criteria. pp. 119–142.

Stephanie Plön, Victor G. Cockcroft, and William P. Froneman. The Natural History and Conservation of Indian Ocean Humpback Dolphins (*Sousa plumbea*) in South African Waters. pp. 143–162.

Salvatore Cerchio, Norbert Andrianarivelo, and Boris Andrianantenaina. Ecology and Conservation Status of Indian Ocean Humpback Dolphins (*Sousa plumbea*) in Madagascar. pp. 163–200.

Muhammad Shoaib Kiani and Koen Van Waerebeek. A Review of the Status of the Indian Ocean Humpback Dolphin *Sousa plumbea* in Pakistan. pp. 201–228.

Dipani Sutaria, Divya Panicker, Ketki Jog, Mihir Sule, Rahul Muralidharan, and Isha Bopardikar. Humpback Dolphins (Genus *Sousa*) in India: An Overview of Status and Conservation Issues. pp. 229–256.

Volume 73, 2016.

Thomas A. Jefferson and Brian D. Smith. Re-assessment of the Conservation Status of the Indo-Pacific Humpback Dolphin (*Sousa chinensis*) Using the IUCN Red List Criteria. pp. 1–26.

Leszek Karczmarski, Shiang-Lin Huang, Carmen K. M. Or, Duan Gui, Stephen C. Y. Chan, Wenzhi Lin, Lindsay Porter, Wai-Ho Wong, Ruiqiang Zheng, Yuen-Wa Ho, Scott Y. S. Chui, Angelico Jose C. Tiongson, Yaqian Mo, Wei-Lun Chang, John H. W. Kwok, Ricky W. K. Tang, Andy T. L. Lee, Sze-Wing Yiu, Mark Keith, Glenn Gailey, and Yuping Wu. Humpback Dolphins in Hong Kong and the Pearl River Delta: Status, Threats and Conservation Challenges. pp. 27–64.

Bernd Würsig, E.C.M. Parsons, Sarah Piwetz, and Lindsay Porter. The Behavioural Ecology of Indo-Pacific Humpback Dolphins in Hong Kong. pp. 65–90.

John Y. Wang, Kimberly N. Riehl, Michelle N. Klein, Shiva Javdan, Jordan M. Hoffman, Sarah Z. Dungan, Lauren E. Dares, and Claryana Araújo-Wang. Biology and Conservation of the Taiwanese Humpback Dolphin, *Sousa chinensis taiwanensis*. pp. 91–118.

Bingyao Chen, Xinrong Xu, Thomas A. Jefferson, Paula A. Olson, Qiurong Qin, Hongke Zhang, Liwen He, and Guang Yang. Conservation Status of the Indo-Pacific Humpback Dolphin (*Sousa chinensis*) in the Northern Beibu Gulf, China. pp. 119–140.

Gianna Minton, Anna Norliza Zulkifli Poh, Cindy Peter, Lindsay Porter, and Danielle Kreb. Indo-Pacific Humpback Dolphins in Borneo: A Review of Current Knowledge with Emphasis on Sarawak. pp. 141–156.

Guido J. Parra and Daniele Cagnazzi. Conservation Status of the Australian Humpback Dolphin (*Sousa sahulensis*) Using the IUCN Red List Criteria. pp. 157–192.

Daniella M. Hanf, Tim Hunt, and Guido J. Parra. Humpback Dolphins of Western Australia: A Review of Current Knowledge and Recommendations for Future Management. pp. 193–218.

Isabel Beasley, Maria Jedensjö, Gede Mahendra Wijaya, Jim Anamiato, Benjamin Kahn, and Danielle Kreb. Observations on Australian Humpback Dolphins (*Sousa sahulensis*) in Waters of the Pacific Islands and New Guinea. pp. 219–272.

Alexander M. Brown, Lars Bejder, Guido J. Parra, Daniele Cagnazzi, Tim Hunt, Jennifer L. Smith, and Simon J. Allen. Sexual Dimorphism and Geographic Variation in Dorsal Fin Features of Australian Humpback Dolphins, *Sousa sahulensis*. pp. 273–314.

Volume 74, 2016.

J. Salinger, A.J. Hobday, R.J. Matear, T.J. O'Kane, J.S. Risbey, P. Dunstan, J.P. Eveson, E.A. Fulton, M. Feng, É.E. Plagányi, E.S. Poloczanska, A.G. Marshall, and P.A. Thompson. Decadal-Scale Forecasting of Climate Drivers for Marine Applications. pp. 1–68.

S.A. Foo and M. Byrne. Acclimatization and Adaptive Capacity of Marine Species in a Changing Ocean. pp. 69–116.

N.D. Gallo and L.A. Levin. Fish Ecology and Evolution in the World's Oxygen Minimum Zones and Implications of Ocean Deoxygenation. pp. 117–198.

R.J. Olson, J.W. Young, F. Ménard, M. Potier, V. Allain, N. Goñi, J.M. Logan, and F. Galván-Magaña. Bioenergetics, Trophic Ecology, and Niche Separation of Tunas. pp. 199–344.

Volume 75, 2016.

G. Notarbartolo di Sciara. Marine Mammals in the Mediterranean Sea: An Overview. pp. 1–36.

L. Rendell and A. Frantzis. Mediterranean Sperm Whales, *Physeter macrocephalus*: The Precarious State of a Lost Tribe. pp. 37–74.

G. Notarbartolo di Sciara, M. Castellote, J.-N. Druon, and S. Panigada. Fin Whales, *Balaenoptera physalus*: At Home in a Changing Mediterranean Sea?. pp. 75–102.

M. Podestà, A. Azzellino, A. Cañadas, A. Frantzis, A. Moulins, M. Rosso, P. Tepsich, and C. Lanfredi. Cuvier's Beaked Whale, *Ziphius cavirostris*, Distribution and Occurrence in the Mediterranean Sea: High-Use Areas and Conservation Threats. pp. 103–140.

R. Esteban, P. Verborgh, P. Gauffier, D. Alarcón, J.M. Salazar-Sierra, J. Giménez, A.D. Foote, and R. de Stephanis. Conservation Status of Killer Whales, *Orcinus orca*, in the Strait of Gibraltar. pp. 141–172.

P. Verborgh, P. Gauffier, R. Esteban, J. Giménez, A. Cañadas, J.M. Salazar-Sierra, and R. de Stephanis. Conservation Status of Long-Finned Pilot Whales, *Globicephala melas*, in the Mediterranean Sea. pp. 173–204.

A. Azzellino, S. Airoldi, S. Gaspari, C. Lanfredi, A. Moulins, M. Podestà, M. Rosso, and P. Tepsich. Risso's Dolphin, *Grampus griseus*, in the Western Ligurian Sea: Trends in Population Size and Habitat Use. pp. 205–232.

D. Kerem, O. Goffman, M. Elasar, N. Hadar, A. Scheinin, and T. Lewis. The Rough-Toothed Dolphin, *Steno bredanensis*, in the Eastern Mediterranean Sea: A Relict Population?. pp. 233–258.

J. Gonzalvo, G. Lauriano, P.S. Hammond, K.A. Viaud-Martinez, M.C. Fossi, A. Natoli, and L. Marsili. The Gulf of Ambracia's Common Bottlenose Dolphins, *Tursiops truncatus*: A Highly Dense and yet Threatened Population. pp. 259–296.

G. Bearzi, S. Bonizzoni, N.L. Santostasi, N.B. Furey, L. Eddy, V.D. Valavanis, and O. Gimenez. Dolphins in a Scaled-Down Mediterranean: The Gulf of Corinth's Odontocetes. pp. 297–332.

M.C. Fontaine. Harbour Porpoises, *Phocoena phocoena*, in the Mediterranean Sea and Adjacent Regions: Biogeographic Relicts of the Last Glacial Period. pp. 333–358.

G. Notarbartolo di Sciara and S. Kotomatas. Are Mediterranean Monk Seals, *Monachus monachus*, Being Left to Save Themselves from Extinction?. pp. 359–386.

T. Scovazzi. The International Legal Framework for Marine Mammal Conservation in the Mediterranean Sea. pp. 387–416.

Volume 76, 2017.

K.S. Meyer. Islands in a Sea of Mud: Insights From Terrestrial Island Theory for Community Assembly on Insular Marine Substrata. pp. 1–40.

E.M. Montgomery, J.-F. Hamel, and A. Mercier. Patterns and Drivers of Egg Pigment Intensity and Colour Diversity in the Ocean: A Meta-Analysis of Phylum Echinodermata. pp. 41–104.

F. Espinosa and G.A. Rivera-Ingraham. Biological Conservation of Giant Limpets: The Implications of Large Size. pp. 105–156.

L. Yebra, T. Kobari, A.R. Sastri, F. Gusmão, and S. Hernández-León. Advances in Biochemical Indices of Zooplankton Production. pp. 157–240.

CHAPTER ONE

Introduction to Northeast Pacific Shark Biology, Ecology, and Conservation

Dayv Lowry*[,1], Shawn E. Larson[†]

*Marine Fish Science Unit, Washington Department of Fish and Wildlife, Olympia, WA, United States
[†]Seattle Aquarium, Seattle, WA, United States
[1]Corresponding author: e-mail address: dayv.lowry@dfw.wa.gov

A Bluntnose Sixgill Shark *Hexanchus griseus* lurks in the depths of Puget Sound, Washington.

Contents

Abstract

Sharks are iconic, sometimes apex, predators found in every ocean and, as a result, they have featured prominently in the mythology, history, and fisheries of diverse human cultures around the world. Because of their regional significance to fisheries and ecological role as predators, and as a result of concern over long-term stability of their populations, there has been an increasing amount of work focused on shark

Advances in Marine Biology, Volume 77
ISSN 0065-2881
http://dx.doi.org/10.1016/bs.amb.2017.06.004

conservation in recent decades. This volume highlights the biodiversity and biological attributes of, and conservation efforts targeted at, populations of sharks that reside in the Northeast Pacific Ocean bordering the west coast of the United States and Canada, one of the most economically and ecologically important oceanic regions in the world. A companion volume addresses details of fisheries and ecotourism in the same region, as well as delving into the relationship between captive husbandry of sharks and education/outreach efforts aimed at fostering a conservation mindset in the public at large. Together, these volumes provide readers a detailed backdrop against which to consider their own actions, and those of resource managers, academics, and educators, as they relate to the long-term conservation of sharks and their relatives.

1. INTRODUCTION

Sharks, and their relatives the skates, rays, and chimeras (collectively making up the class Chondrichthyes), are found throughout every ocean in the world and in some freshwater lakes and river systems (Compagno, 2001). Some species, such as the Bull Shark, *Carcharhinus leucas*, and the river shark, genus *Glyphis*, can even thrive in both saltwater and freshwater. In marine environments, they can be found from the shallows of nearshore estuaries, to the pelagic environments of the open ocean, to the depths of abyssal plains. Chondrichthyan fishes have skeletal structures made of cartilage, a relatively hard but flexible tissue, rather than calcified bones making them very flexible and agile in the water. The diversity of body forms, behaviours, and ecological roles exhibited by sharks has evolved over millennia to make them well suited to thrive in diverse and dynamic habitats. As vertebrate fishes that have inhabited the oceans for hundreds of millions of years, they are among the most ancient extant lineages of fishes, along with the jawless Agnathans (lamprey and hagfish) and lobe-finned fishes such as coelacanth (*Latimeria* spp.), the oldest "living fossil" known (Compagno, 2001; Forey, 1998; Janvier, 2010).

There are currently over 500 species of sharks recognized by science, with new species discovered and classified by taxonomists nearly every year. They range in size from the Dwarf Lanternshark (*Etmopterus perryi*), a deep-sea species reaching only 17 cm in total length, to the Whale Shark (*Rhincodon typus*), the largest fish in the world at approximately 12 m long (Compagno, 2001). Many large (>2 m total length) sharks are higher-order predators and keystone species that shape the ecosystems in which they live; however, most shark species occupy a more modest position in the food web as predators of small fishes and crustaceans, and prey for larger fishes and marine mammals. The combination of intense localized fishing and life histories characterized by long lives, slow maturation, and low fecundity makes many chondrichthyan species particularly vulnerable to population decline at the hands of humans (Musick, 1999).

Recent estimates are that a quarter of all elasmobranch species (sharks, skates, and rays, but not chimeras) are threatened with extinction, and only 25% of the 524 known shark species are categorized as being of Least Concern on the IUCN Red List (Dulvy et al., 2014). The effects of removing sharks from our aquatic ecosystems are complex and varied, and quantitatively predicting impacts of such removals requires data that are often in short, but ever expanding, supply (Heithaus et al., 2008). However, we know from other systems that the removal of apex predators, such as large carnivores (Ripple et al., 2014), can cause diverse changes in prey behaviour and survival rates that result in trophic cascades or collapses of trophic levels below the apex predator, disrupting ecosystem functioning and biodiversity (e.g., Borer et al., 2005; Pace et al., 1999; Schmitz et al., 1997). In marine ecosystems, this can also lead to the potential loss of economically important fisheries as predatory pressure is relieved on prey species that, in turn, consume species upon which fisheries depend (Myers et al., 2007).

This volume highlights the biological attributes of, and the conservation efforts targeted at, populations of sharks residing in the Northeast Pacific Ocean (NEP) bordering the west coast of the United States and Canada, one of the most economically and ecologically important oceanic regions in the world (Center for Ocean Solutions, 2009; FAO, 2009). The NEP, here, is defined as the Pacific Ocean from the Bering Strait in the north, down the west coast of North America including the Aleutian Archipelago, the Gulf of Alaska, and the nearshore (inside) waters of the Alexander Archipelago; the Straits of Juan de Fuca and Puget Sound; San Francisco Bay, the Southern California Bight; the Gulf of California (or Sea of Cortez); the waters surrounding Baja California; and the waters along the west coast of Mexico to its border with Guatemala (Fig. 1). In some instances, chapters in this volume, and its companion Volume 78 of *Advances in Marine Biology* (AMB), limit their focus to only a portion of this region based on regionally available information or for heuristic/demonstrative purposes when addressing broad topics.

The impetus for this book was born out of local, biennial shark conservation meetings held at the Seattle Aquarium starting in 2004 (http://www.seattleaquarium.org/shark-workshop). The first of these conservation workshops focused on a single taxon, the cowsharks of family Hexanchidae, specifically Bluntnose Sixgill (*Hexanchus griseus*) and Broadnose Sevengill (*Notorynchus cepedianus*) sharks. These meetings brought together partners in six- and sevengill research on the west coast of North America from

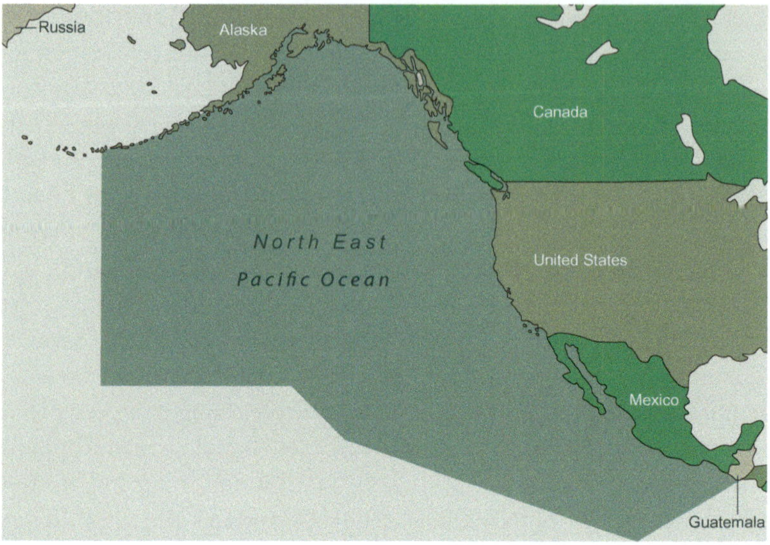

Fig. 1 Map of the Northeast Pacific Ocean (NEP) as defined for use in this volume and its companion volume.

California to Vancouver, BC. The Cowshark Conservation Workshops grew in breadth and scope until there was interest in the community to include other shark species. In 2014, for the conservation workshop's 10th anniversary, the biennial meeting evolved to become the Northeast Pacific Shark Symposium (http://www.seattleaquarium.org/shark-workshop). The topics and new knowledge presented at these symposia stimulated interest in putting together a book dedicated to the conservation biology and ecology of NEP sharks such that a broader audience could be made aware of recent developments in fields ranging from fundamental taxonomy and genetic examination of population connectivity, to population modelling and fisheries policy, to shark-focused ecotourism. As plans further developed, it was determined that two volumes would be produced: the first dealing with systematics and aspects of organismal biology and growth (AMB Volume 77), and the second focusing on fisheries, captive husbandry, and other population-level elements of human interactions with sharks (AMB Volume 78).

2. VOLUME 77 CONTENT

The chapters of this volume, and its companion, are arranged to provide readers a comprehensive understanding of the current state of the knowledge of sharks occurring in the NEP, from the fundamentals of taxonomy through advanced population modelling and resource management. A dedicated chapter on management and policy is not provided; however, each chapter references diverse regional, national, and international policy documents and actions as germane. Among these are well-known documents such as the Endangered Species Act (ESA), the Convention on International Trade in Endangered Species (CITES), and International Union for Conservation of Nature (IUCN) Red List assessments. In addition, elements of state and provincial management regimes that impact sharks are discussed, and considerable detail is provided, when appropriate, regarding management of marine resources by the Pacific Fishery Management Council and the North Pacific Fishery Management Council under the auspices of the Magnuson-Stevens Fishery Conservation and Management Act (National Marine Fisheries Service, 2007).

In Chapter 2, renowned elasmobranch taxonomist and researcher Dr. Dave Ebert of the Pacific Shark Research Center (PSRC) at the Moss Landing Marine Laboratories (MLML), along with Jenny Bigman and IUCN Shark Specialist Group Program Officer Julia Lawson of Simon Fraser University, delves into the biodiversity of sharks known to occur in the NEP. Far from presenting merely a laundry list of names and geographic distributions, the authors also discuss the critical role that systematics and taxonomy play in identifying and geographically delineating species, and stocks, of sharks such that targeted management actions can act on biologically appropriate units. Without a thorough understanding of variations in species-specific life history and distribution, as well as species composition of the predator complex, many conservation actions fall short of achieving their desired goals simply due to a lack of understanding of the fundamental factors that drive population change.

In addition to using traditional descriptions of body element shape/size (i.e., morphometrics) and tallies (i.e., meristics), a variety of genetic analyses can be used to distinguish species, and demographically independent subunits of species, such that management and conservation efforts can be more finely tailored. Several variants of genetic analyses focused on DNA extracted from various portions of cells can be used to infer population boundaries, male and female dispersal, parentage, mating systems, and relatedness of individuals captured in fisheries or observed by survey methods. In addition to reviewing how these conservation genetics approaches have been applied to NEP shark species over the past decade, Dr. Shawn E. Larson of the Seattle Aquarium, Dr. Toby Daly-Engel of Florida Institute of Technology, and Dr. Nicole Phillips of University of Southern Mississippi also explore how recently devised sampling techniques, such as the collection of environmental DNA (eDNA), might be employed to guide long-term conservation actions in the future.

Building upon an understanding of the constituent species of the shark complex occupying the NEP and the tools used to delineate them and probe into their demographic processes, Dr. Joseph Bizzarro of the PSRC and National Marine Fisheries Service (NMFS), Dr. Aaron Carlisle of the Hopkins Marine Station of Stanford University, Dr. Wade Smith of the University of British Columbia, and Dr. Enric Cortes from the NMFS' Southeast Fisheries Science Center summarize what is known about the ecological role of these predators. Far from all sharks conforming to the large, pelagic, apex predator mould often associated with Great White Sharks (*Carcharodon carcharias*) and their relatives, the authors demonstrate that most of the shark species found in the region are in fact rather small (<2 m total length), nearshore, mid-tier predators that have nearly as many predators as they do prey species. Corresponding with extensive diversity in size, morphology, predatory behaviour, and habitat, the authors document broad-scale diversity in diet and consumption rates, both of which are influenced by myriad factors such as age, sex, maturation stage, water temperature, and season. The authors also discuss the ecosystem-scale impacts of variation in diet and ecological interactions, which range well beyond who eats whom, and can influence community structure and spatial utilization of habitat by prey species and competitors on both the small and large scale. Finally, the authors note the critical lack of trophic models, necessary for many effective conservation efforts, which include impacts of sharks and other predators for many regions of the NEP.

While consumption of prey and interactions with competitors and predators are key elements of the ecology of both individuals and shark populations as a whole, understanding how diverse prey resources are integrated into the bodies of consumers through metabolic processes is also critically important to incorporate into management and conservation actions. In their chapter, Drs. Jonathan Reum and Chris Harvey, and Mr. Greg Williams, of the NMFS' Northwest Fisheries Science Center review how stable isotope analysis can be used to directly quantify the integration of specific suites of organic material from various food chains into the tissues of sharks, providing information about their trophic level and feeding habits that cannot be gleaned from analyses of gut contents alone. When specific isotopic signatures can be identified for distinct food chains over time and through space, such as offshore pelagic and nearshore benthic, this information can also be used to infer geographic movement patterns of individuals. In cases where individuals of a species occupy deep, offshore, or other areas that are logistically difficult and expensive to sample, these isotopic analyses can provide an integrated snapshot of ecological interactions that would otherwise be unattainable. Accounting for variation in such movement patterns, and incorporating this into population movement models, can help elucidate the placement of a species within a food web and better allow for the conservation of resources upon which the species relies.

When evaluating the abundance, distribution, and stock structure of a species for management and conservation, it is extremely helpful to have knowledge of the way in which individual age classes contribute to these metrics. Fundamental to being able to assess these contributions is the ability to confidently assign individuals an age. Sharks lack otoliths (calcified ear bones), and for some species, especially those in which skeletal calcification is minimal, other traditional materials used to age bony fishes, such as vertebral centra and neural arches, are minimally informative and techniques have been difficult to replicate. For some species, vertebral centra or neural arches can be used successfully, but the deposition of growth bands can be annual or biannual. For still other species (e.g., the North Pacific Spiny Dogfish *Squalus suckleyi*), nonskeletal hard parts, specifically keratinous dorsal spines, are the most reliable indicator of age. In their chapter, NOAA Fisheries research biologists Beth Matta, Dr. Cindy Tribuzio, and Chris Gburski, along with Dr. Dave Ebert of the PSRC and MLML and Dr. Ken Goldman from the Alaska Department of Fish and Game, review the idiosyncrasies associated with ageing those species for which an ageing technique has been identified and also discuss the application of novel methods for species without an established method. The authors also discuss the importance of accurate ageing when it comes to modelling population parameters for fisheries.

With species diversity identified, the ecology of many species described, and methods for establishing many of the key data elements needed to model population dynamics for fishery management established, Volume 77 ends thus setting the stage for Volume 78 where we explore how humans interact with sharks both in positive and negative ways, such as via fisheries, captive husbandry, and ecotourism. The focus of this second volume will be similar to the current volume, highlighting the conservation implications of human interactions on sharks within the NEP.

REFERENCES

Borer, E.T., Seabloom, E.W., Shurin, J.B., Anderson, K.E., Blanchette, C.A., Broitman, B., Cooper, S.D., Halpern, B.S., 2005. What determines the strength of a trophic cascade? Ecology 86 (2), 528–537.

Center for Ocean Solutions, 2009. Pacific Ocean Synthesis: Scientific Literature Review of Coastal and Ocean Threats, Impacts, and Solutions. The Woods Center for the Environment, Stanford University, CA. 158 pp.

Compagno, L.J.V., 2001. Sharks of the world: an annotated and illustrated catalogue of shark species known to date. In: Bullhead, Mackerel and Carpet Sharks (Heterodontiformes, Lamniformes and Orectolobiformes). FAO Species Catalogue for Fishery Purposes, No. 1, vol. 2. FAO, Rome, Italy.

Dulvy, N.K., Fowler, S.L., Musick, J.A., Cavanagh, R.D., Kyne, P.M., Harrison, L.R., Carlson, J.K., Davidson, L.N.K., Fordham, D.V., Francis, M.P., Pollock, C.M., Simpfendorfer, C.A., Burgess, G.H., Carpenter, K.E., Compagno, L.J.V., Ebert, D.A., Gibson, C., Heupel, M.R., Livingstone, S.R., Sanciangco, J.C., Stevens, J.D., Valenti, S., White, W.T., 2014. Extinction risk and conservation of the world's sharks and rays. eLife 3, e00590.

FAO Fisheries and Aquaculture Department, 2009. The State of World Fisheries and Aquaculture: 2008. Food and Agriculture Organization of the United Nations, Rome, Italy. 196 pp.

Forey, P., 1998. A home from home for coelacanths. Nature 395 (6700), 319–320.

Heithaus, M.R., Frid, A., Wirsing, A.J., Worm, B., 2008. Predicting ecological consequences of marine top predator declines. Trends Ecol. Evol. 23 (4), 202–210.

Janvier, P., 2010. microRNAs revive old views about jawless vertebrate divergence and evolution. Proc. Natl. Acad. Sci. U. S. A. 107 (45), 19137–19138.

Musick, J.A., 1999. Criteria to define extinction risk in marine fishes: the American Fisheries Society initiative. Fisheries 24 (12), 6–14.

Myers, R.A., Baum, J.K., Shepherd, T.D., Powers, S.P., Peterson, C.H., 2007. Cascading effects of the loss of apex predatory sharks from a coastal ocean. Science 315 (5820), 1846–1850.

National Marine Fisheries Service, 2007. Magnuson-Stevens Fishery Conservation and Management Act: As Amended Through January 12, 2007. U.S. Department of Commerce, National Oceanic and Atmospheric Administration. 170 pp.

Pace, M.L., Cole, J.J., Carpenter, S.R., Kitchell, J.F., 1999. Trophic cascades revealed in diverse ecosystems. Trends Ecol. Evol. 14 (12), 483–488.

Ripple, W.J., Estes, J.A., Beschta, R.L., Wilmers, C.C., Ritchie, E.G., Hebblewhite, M., Berger, J., Elmhagen, B., Letnic, M., Nelson, M.P., Schmitz, O.J., Smith, D.W., Wallach, A.D., Wirsing, A.J., 2014. Status and ecological effects of the world's largest carnivores. Science 343 (6167), 1241484.

Schmitz, O.J., Beckerman, A.P., O'Brien, K.M., 1997. Behaviorally mediated trophic cascades: effects of predation risk on food web interactions. Ecology 78, 1388–1399.

CHAPTER TWO

Biodiversity, Life History, and Conservation of Northeastern Pacific Chondrichthyans

David A. Ebert*,[1], Jennifer S. Bigman†, Julia M. Lawson†
*Pacific Shark Research Center, Moss Landing Marine Laboratories, Moss Landing, CA, United States
†Earth to Ocean Research Group, Simon Fraser University, Burnaby, BC, Canada
[1]Corresponding author: e-mail address: debert@mlml.calstate.edu

Eastern Pacific Black Ghostshark (*Hydrolagus melanophasma*). AKA Ratfish or chimaera. This fish was photographed at a depth of about 1750 m in the Guymas Basin of the Gulf of California. (c) 2003 MBARI.

Contents

Advances in Marine Biology, Volume 77
ISSN 0065-2881
http://dx.doi.org/10.1016/bs.amb.2017.07.001

Abstract

The sharks, batoids, and chimaeras, collectively the class Chondrichthyes, are one of the most successful groups of fishes, with over 1250 species globally. Recent taxonomic revisions have increased their diversity by about 20% over the past 17 years (2000–2016). The Northeast Pacific Ocean is one of the top 20 most diverse regions/countries on the globe with 77 chondrichthyan species, a number less than a quarter that of the most species-rich area (Australia) but that has increased by 10% since 2000 to include three new species (two skates and a chimaera). In this chapter we discuss the species richness of chondrichthyans occurring in the Northeast Pacific Ocean, characterize their life histories, briefly review several fisheries, and summarize the conservation status of those chondrichthyans occurring in the region. Detailed descriptions and evaluations of fisheries can be found in Chapter 7 of AMB Volume 78.

1. INTRODUCTION

Sharks and their relatives the batoids and chimaeras, collectively comprising the class Chondrichthyes, are one of the most successful groups of fishes, having penetrated most marine ecosystems (Compagno, 1990; Nelson et al., 2016). These ecosystems include tropical and high-latitude coastal, continental, and insular shelves, and extend into the deep sea (>200 meters (m)) as well as the open ocean. In some areas, certain species either reside in or seasonally utilize estuarine and freshwater river ecosystems. The diversity of known species is highest in tropical marine ecosystems, followed by the deep-sea and temperate ecosystems, with pelagic and freshwater ecosystems having lower diversity. The majority of species (~55%) occur on continental shelves from the intertidal zone, including bays and estuaries, along coastal nearshore areas and out to about 200 m depth (Compagno, 1990; Ebert and Winton, 2010). The diversity of shelf species is greater in the tropics (~40%) and lower in temperate (~30%) seas, while nearly one-half (~50%) of all species are considered deep sea (Ebert and Winton, 2010; Kyne and Simpfendorfer, 2010; White and Sommerville, 2010). Species diversity in pelagic and freshwater habitats is comparatively low, making up only approximately 3% and 2%, respectively (Dulvy et al., 2014a; Stevens, 2010).

Globally, there are approximately 524 known species of sharks, nearly 700 batoid species, and more than 50 chimaera species, bringing the overall total to over 1250 species of sharks and shark-like fishes (Table 1). However, the public's general perception of sharks is typically limited to the

Table 1 Known Biodiversity of Chondrichthyans by Order, Family, Genus, and Species (D.A. Ebert, pers. database, 31 Dec. 2016)

Order	Family	(%)	Genera	(%)	Species	(%)
Hexanchiformes (Cow and Frilled Sharks)	2	3.2	4	1.8	7	0.5
Echinorhiniformes (Bramble Sharks)	1	1.6	1	0.5	2	0.1
Squaliformes (Dogfish Sharks)	6	9.5	23	10.6	130	9.2
Pristiophoriformes (Sawsharks)	1	1.6	2	0.9	8	0.6
Squatiniformes (Angel Sharks)	1	1.6	1	0.5	21	1.5
Heterodontiformes (Horn or Bullhead Sharks)	1	1.6	1	0.5	9	0.6
Orectolobiformes (Carpet Sharks)	7	11.1	14	6.5	46	3.2
Lamniformes (Mackerel Sharks)	7	11.1	10	4.6	15	1.1
Carcharhiniformes (Ground Sharks)	8	12.7	51	23.5	286	20.1
Torpediniformes (Electric, Thornback, and Fan Rays)	5	7.9	15	6.9	62	4.4
Rhinopristiformes (Sawfish, Guitarfish, and Wedgefish)	5	7.9	11	5.1	65	4.6
Rajiformes (Skates)	4	6.3	38	17.5	298	21.0
Myliobatiformes (Devilray, Eagleray, and Stingray)	12	19.0	40	18.4	262	18.5
Chimaeriformes (Chimaeras, Ghost Sharks, Ratfish, or Silver Sharks)	3	4.8	6	2.8	51	3.6
Total	63		217		1262	

more iconic forms, e.g., White Shark (*Carcharodon carcharias*), Tiger Shark (*Galeocerdo cuvier*), or other distinct species of large, fearsome, and toothy predators. The reality is that sharks come in a variety of sizes and shapes from the Whale Shark (*Rhincodon typus*), the world's largest fish, to the Pygmy Shark (*Euprotomicrus bispinatus*), one of the smallest fish species. Only 18% of known shark species have known maximum lengths exceeding 2m or known maximum weights exceeding 10kg (Compagno et al., 1989). While they are all predatory, they also occupy a variety of niches in the food web and feed on lower trophic tier prey ranging from plankton, in the case of the Whale Shark and Basking Shark (*Cetorhinus maximus*), to diverse invertebrates and small fishes (Compagno, 1990; see chapter "Diet composition and trophic ecology of Northeast Pacific sharks" by Bizzarro et al., this volume).

The known diversity of sharks and their relatives, hereafter collectively referred to as sharks, has increased exponentially over the past 17 years (2000–2016) with approximately 280 new species having been described (Fig. 1). This represents over 20% of all valid shark species that have been described to date, which compares to the approximately 330 species that had been described over the previous 50 years (1950–1999). The majority of recent new species discoveries have come mostly from the

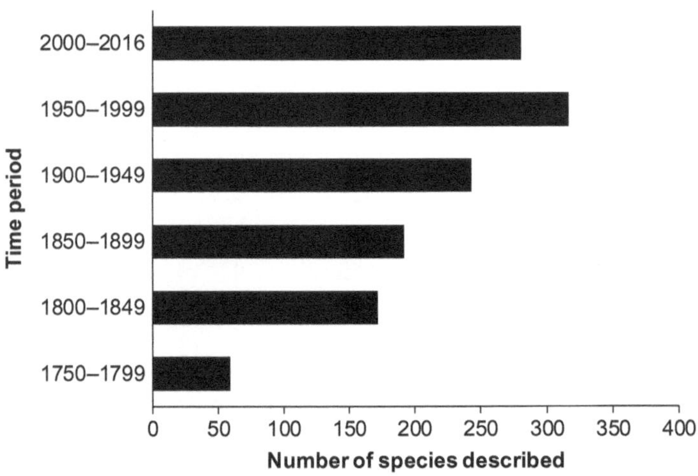

Fig. 1 The number of newly described species of sharks and other chondrichthyes over time since 1750. Counts are separated into 50-year intervals except for the most recent increment (2000–2016). Despite being only one-third the duration, the number of newly described species in this latest period is greater than any other period except 1950–1999.

Indo-Australian region, followed by the southern African and western North Pacific regions (Ebert and van Hees, 2015; Ebert et al., 2013a).

The purpose of this chapter is to highlight the species richness of chondrichthyans occurring in the Northeast Pacific Ocean (NEP), characterize their life histories, briefly review several fisheries, and summarize their conservation status.

2. NEP ECOSYSTEM

The distribution of marine organisms is influenced by a variety of factors, with temperature being the principle factor determining many species' geographic ranges. Most marine species exhibit horizontally stratified distribution patterns, with species diversity decreasing from warm equatorial to cool high-latitude seas (Briggs, 1995). Species distribution patterns and the latitudinal thermal gradient can be used to subdivide the ocean into four zones: tropical, warm-temperate, cold-temperate, and polar zones.

The NEP region, also referred to as the Eastern North Pacific, as considered here, extends from approximately 65°N, 180°W to about 32°N, 117°W or from just south of the Bering Strait eastwards along the Eastern Bering Sea continental slope to the Gulf of Alaska and southwards along the continental landmass of North America to the California, USA/Mexico border. The majority of the NEP is categorized as a cold-temperate region that extends along the North American Pacific coast from the eastern Bering Sea southwards to Point Conception, California. The extent of the cold-temperate zone in this region is determined primarily by the eastward flowing North Pacific Current, which splits to form the California Current to the south and the Alaska Current to the north. The major change from a cold- to warm-temperate water regime occurs at Point Conception (~34.4°N, ~120.5°W), California. The NEP encompasses three large marine ecosystems: the Eastern Bering Sea and Aleutian Islands (EBS), the Gulf of Alaska (GOA), and the California Current (CAC). While the CAC extends from the Canada/USA border to the tip of Baja California, and includes the Gulf of California, here we only consider that portion extending to southern California to be the most southerly point in the NEP. For fisheries purposes, the Food and Agricultural Organization of the United Nations (FAO) divides the NEP, as defined here, into two Major Fishing Areas (MFAs). Major Fishing Area 67 is named the Pacific, Northeast and encompasses the waters from the most easterly point of mainland Russia to about Cape Mendocino, California, USA. Major Fishing Area

77 is named Pacific, Eastern Central and encompasses the remainder of California, as well as the west coasts of Mexico and Central America (to the Panama–Colombia border). The NEP chondrichthyan fauna therefore mostly inhabits a boreal cold-temperate regime, but this shifts to a warm-temperate regime in the southern reaches of the region.

3. CLASSIFICATION

The chondrichthyans, especially the batoids, are currently undergoing extensive revisions at higher classification levels, with a reexamination of the phylogenetic relationships of these various groups (Aschliman et al., 2012; Last et al., 2016; Naylor et al., 2012a,b). The relationship of the sharks to the batoids is unresolved, with recent classification schemes suggesting that the batoids are either sister to the modern shark orders Pristiophoriformes and Squatiniformes, and share a common ancestry with the Squaliformes (i.e. the Hypnosqualean hypothesis), or follow the traditional dichotomy of all modern sharks and batoids. Evidence for the former hypothesis comes largely from newer molecular evidence (Aschliman et al., 2012; Douady et al., 2003; Naylor et al., 2005, 2012a,b), while traditional morphological evidence (Compagno, 1973, 1977, 2001, 2005; de Carvalho, 1996; Shirai, 1992, 1996) has been, and continues to be, the cornerstone of the latter phylogenetic classification.

The class Chondrichthyes has two major lineages, the subclass Neoselachii, which is subdivided into two cohorts, the Selachii (sharks) and the Batoidea (rays), and the subclass Holocephali, which includes all chimaeras in a single order Chimaeriformes (Nelson et al., 2016). The Selachii is subdivided into two superorders, Squalomorphii and Galeomorphii, comprised of nine orders. The Squalomorphii includes five orders (Hexanchiformes, Echinorhiniformes, Squaliformes, Squatiniformes, and Pristiophoriformes) and the superorder Galeomorphii includes four orders (Heterodontiformes, Orectolobiformes, Lamniformes, and Carcharhiniformes). The cohort Batoidea includes four orders, Torpediniformes, Rhinopristioformes, Rajiformes, and Myliobatiformes (Last et al., 2016). The higher classification, i.e., order, family, and genera, used here for the sharks follows Ebert (2016), Ebert et al. (2013b), and Nelson et al. (2016); for the batoids follows Last et al. (2016) and Eschmeyer et al. (2016); and for the chimaeras follows Didier et al. (2012). Some additional classification modifications were included where appropriate to reflect more recent changes, based largely on recent genetic studies.

4. ZOOGEOGRAPHY

The checklist of species provided here was compiled from individual species checklists, regional catalogues and guides, literature accounts, museum voucher specimens, and original data gathered by the authors. The references include mostly selected major or significant works in the region. This included Ebert (2003), Eschmeyer et al. (2016), Hart (1973), Mecklenburg et al. (2002), and various individual International Union for the Conservation of Nature (IUCN) Red List of Threatened Species assessments.

The NEP region is among the top 20 most diverse regions, as defined by the Food and Agriculture Organization, for chondrichthyans in the world with 77 species (Table 2). The top three regions with the greatest chondrichthyan diversity, Australia, India, and Japan, by comparison have 323, 227, and 212 chondrichthyan species, respectively. The NEP region, however, is well represented at higher taxonomic levels with 12 orders, 33 families, and 49 genera present; two genera, *Isistius* and *Zameus*, have been added since 2014 (Ebert et al., 2014, 2015). Of the 9 recognized shark orders, 8 are present in this area, with 20 families, 32 genera, and 42 species confirmed from the region. The batoids have 3 orders, 11 families, 15 genera, and 31 species represented, while the chimaera order has 2 families, 2 genera, and 4 species represented. A checklist of current species is provided in Appendix A.

Diversity of the fauna within the NEP is relatively uneven, with the number of species occurring across the three large marine ecosystems (EBS, GOA, CAC) increasing noticeably to the south (Table 3). The cold-temperate EBS is least diverse with only 3 orders, 5 families, 7 genera, and 18 species, mostly skates. The skates, with 15 species compared to only 3 shark species, dominate the EBS chondrichthyan fauna; no chimaera species are known in this ecosystem. The GOA is slightly more diverse, with 9 orders, 17 families, 23 genera, and 33 species. However, this is likely due to the influence of the northward flow of the Alaska Current into southeast Alaska, which occasionally brings uncommon, oceanic species, such as the Bigeye Thresher Shark (*Alopias superciliosus*) and the Pelagic Stingray (*Pteroplatytrygon violacea*), into the southern reaches of this ecosystem. The number of shark and batoid species is similar with 16 species each, but 9 sharks compared to 5 batoid species are uncommon or considered vagrants in this ecosystem. The skates are the most dominant group with 13 species.

Table 2 Top 20 Regions/Countries With High Biodiversity of Sharks (D.A. Ebert, pers. database)

Region	Total	Shark	Batoid	Chimaera
Australia	323	183	123	15
India	227	114	107	6
Japan	212	126	75	11
Southern Africa[a]	205	118	79	8
Indonesia	192	103	86	5
Western Central Atlantic[b]	188	103	79	6
Taiwan	181	119	58	4
Eastern Central Pacific[c]	176	88	88	8
Brazil	169	82	81	6
Southeastern Pacific[d]	165	86	71	8
Arabian Sea[e]	142	73	69	2
Philippines	165	95	67	3
Northeast Atlantic[f]	127	70	49	8
New Zealand	119	69	33	17
Mexico (Pacific Coast)	113	58	51	4
Argentina	110	55	52	2
Mediterranean	83	49	33	1
Northwest Atlantic[g]	83	54	23	6
Red Sea	77	44	33	0
Northeast Pacific[h]	77	43	30	4

[a]Southern Africa here includes Angola, South Africa, Mozambique.
[b]Western Central Atlantic includes the Mexican portion of the Gulf of Mexico and Caribbean Sea.
[c]Eastern Central Pacific or tropical eastern Pacific here includes Central America to Ecuador.
[d]Southeastern Pacific includes Peru and Chile.
[e]Arabian Sea includes the Arabian/Persian Gulf, but excludes the Red Sea.
[f]Northeast Atlantic includes European States bordering on the Atlantic Ocean.
[g]Northwest Atlantic includes the Atlantic coast of Canada and the United States.
[h]Northeast Pacific includes the Pacific coast of Canada and the United States.

Table 3 Known Biodiversity of Chondrichthyans Within the NEP by Order, Family, Genus, and Species (D.A. Ebert, pers. database, 31 Dec. 2016)

	Orders	Family	Genera	Species
Northeast Pacific Fauna				
Sharks	8	20	32	42
Batoids	3	11	15	31
Chimaeras	1	2	2	4
Total	12	33	49	77
California Current				
Sharks	8	20	32	42
Batoids	3	11	15	22
Chimaeras	1	2	2	4
Total	12	33	49	68
Gulf of Alaska				
Sharks	5	11	15	16
Batoids	3	5	7	16
Chimaeras	1	1	1	1
Total	9	17	23	33
Eastern Bering Sea				
Sharks	2	3	3	3
Batoids	1	2	4	15
Chimaeras	0	0	0	0
Total	3	5	7	18

The CAC is the most diverse with 68 species, with all NEP orders, families, and genera represented (Ebert, 2003; Ebert et al., 2014, 2015; Reichert et al., 2016). The higher diversity is due to it being the most dynamic of the three large marine ecosystems, with an influx of warm-temperate water, especially off southern California. The south-flowing CAC veers offshore around Point Conception, allowing the warm-temperate waters of the north-flowing Davidson Current to influence the southern California Bight and Channel Islands. The number of families, genera, and species increases from northern to southern California, with the latter being a transitional zone from a cold- to warm-temperate regime. The fauna south of Point Conception is remarkably similar to the Mexican Pacific coast fauna at the family level, with all but two shark and three batoid families being found in both regions (Ebert, 2003). Conversely, north of Point Conception the fauna is similar to the Alaskan fauna, with most families also represented in the GOA. The most significant changes are in the species composition found in each location. The skates in particular become the prominent batoid group north of Point Conception with the Rhinopristiformes and Myliobatiformes becoming scare to absent the further north one goes.

Given the similarity in water temperature and influence of the Davidson Current in the southern CAC, the distribution of several species ranges from southern California, USA, to the tip of Baja California and includes the Gulf of California, Mexico. As such, some species considered endemic to the NEP have at least part of their distribution outside of NEP waters (southwards from Point Conception, California, USA). The IUCN considers endemic species to be those that are found in any specific area and nowhere else (IUCN, 2012). Here, species that had part of their distributions within the NEP (north of Port Conception, California, USA) were considered endemic to the region, even if the majority of their range fell outside of NEP waters, making them endemic to both occupied regions.

5. HISTORY OF SPECIES DISCOVERY

The first chondrichthyan species named from the NEP was the White-spotted Chimaera *Hydrolagus colliei* (Lay and Bennett, 1839), named after the naturalist Alexander Collie, who collected the type specimen from Monterey Bay, California, during the voyage of the *HMS Blossom* (1825–1828) under Captain Beechey that explored the Pacific Northwest as far north as Point Barrow, Alaska. Descriptions of new Chondrichthyan species from the NEP, however, began in earnest with Charles Girard of the Smithsonian Institution in the 1850s. He described four new chondrichthyan species, which are still valid today, from specimens collected during the Pacific Railroad Surveys (1853–1855). However, one of the first resident ichthyologists, a physician named William Ayres who came west in the mid 1800s during the California gold rush, named three species that are still valid Ayres was one of the founders of the California Academy of Sciences.

Ichthyology in North America, and especially the NEP, entered a golden age with the arrival of David Starr Jordan who was, and still is to this day, the dominant figure in the field of ichthyology with most North American researchers tracing their ichthyological roots back to him. The other dominant ichthyological figure from this era was Charles Gilbert, who was Jordan's student and frequent collaborator. Jordan, who later became the founding president of Stanford University, appointed Gilbert as the chairman of the zoology department. Together Jordan and Gilbert named six NEP chondrichthyan species, but Gilbert later went on to describe five more NEP species himself, the most of any researcher in the NEP. Furthermore, two additional species described by Jordan and Gilbert from Mexican waters were later found to also occur within the NEP, giving these collaborators eight and 13 species, respectively, found within the NEP. Theodore Gill was the other notable ichthyologist from this period who named four new chondrichthyan species. Given the intense interest in systematic ichthyology at the time, and with so many prominent ichthyologists, it is not surprising that 26 of 33 species described from the NEP were named during the 19th century, with five in the 20th century, and two so far in the 21st century.

The holotype locations for 33 of 77 known chondrichthyan species are from within the NEP, while the type location for another five species was in waters adjacent to the region. The majority of NEP shark species were described from outside the region, with only nine of 42 species having their type locations within the region. This is likely due to the wide-ranging distribution of most species, many of which have a global distribution. Conversely, the majority of batoids, 23 of 31 species, were described based on type locations within the NEP; this is likely due to the more restricted distribution of most batoid species. One of four chimaera species was described from the NEP; two of the chimaera species have restricted distributions in the eastern Pacific, while two others are wide ranging.

The total number of species now known to occur within the NEP has increased by 10% since 2000, with the addition of at least seven species to the fauna, including the descriptions of three new species: Butterfly Skate, *Bathyraja mariposa*, Leopard Skate, *Bathyraja panthera*, and Eastern Pacific Black Ghost Shark, *Hydrolagus melanophasma* (James et al., 2009; Orr et al., 2011; Stevenson et al., 2004). Two warm-temperate to tropical species, the Longfin Mako Shark *Isurus paucus* and Velvet Dogfish *Zameus squamulosus*, were previously almost entirely unknown from the eastern Pacific. In fact, the only confirmed records of *I. paucus* from the eastern Pacific are from off southern California and Baja California, while *Z. squamulosus* has only been reported from southern California and from the southeastern Pacific (Ebert, 2001, 2016; Ebert et al., 2014). The

Cookiecutter Shark *Isistius brasiliensis* was long suspected to occur within the NEP region based on crater wounds on Northern Elephant Seals *Mirounga angustirostris*, but was only recently confirmed (Ebert et al., 2015). Also, recently added to the fauna is the tentatively identified Pointy-nosed Blue Ghost Shark *Hydrolagus* cf. *trolli*, which has been observed only by remotely operated vehicles from very deep water (>2000 m) around seamounts off central and southern California (Reichert et al., 2016). At least two species (the Bull Shark *Carcharhinus leucas* and the Blacktip Shark *C. limbatus*) may occur within this region, but their distribution in the NEP is uncertain (Ebert, 2003). Both are well known from Baja California, MX waters and may occur off southern California on occasion, but have not yet been confirmed (Ebert, 2003). Resolution of species complexes, such as the *Bathyraja interrupta–kincaidii* complex, may further lead to the identification of additional, possibly cryptic, species within the region, thus increasing overall recognized species diversity.

6. NEP BIODIVERSITY

The Hexanchiformes (Cow and Frilled Sharks) comprises two families, four genera, and seven species of morphologically distinct sharks, in that they are the only group to possess six or seven paired gill slits, a single dorsal fin, and an anal fin. Members of this group of moderately sized to very large sharks mostly occupy a benthopelagic habitat. Both families and three species occur in the NEP. The Chlamydoselachidae are a wide-ranging family, with a single genus and two species, but in the NEP there is only a single record of the wide-ranging Frilled Shark *Chlamydoselachus anguineus* captured off Point Arguello, California. The Hexanchidae, with three genera and five species, are also wide ranging and are represented in the NEP by two very common species. The Broadnose Sevengill Shark *Notorynchus cepedianus* is primarily a coastal, continental shelf species and is very common seasonally in northern California bays, while the Bluntnose Sixgill Shark *Hexanchus griseus* is a very common deep-water species throughout the region.

The Echinorhiniformes (Bramble Sharks), formerly placed in the order Squaliformes, are a small group consisting of a single genus with two wide-ranging, deep-sea species with a scattered, worldwide distribution. The group is morphologically characterized as being heavy bodied, moderate to large in size, and covered with dermal denticles ranging from coarse and spiky to large tack-like thorns. Additionally species have a short-snout, flattened head, no anal fin, and two dorsal fins that are small, spineless, and posteriorly set. A single species, the Prickly Shark *Echinorhinus cookei* is spottily distributed in the NEP from Oregon to southern California, but is relatively common in central California, especially in and around Monterey Bay.

The Squaliformes (Dogfish Sharks) are the third largest group of chondrichthyans with six families, 23 genera, and at least 130 described species. The squaloids are a morphologically diverse group that contains some of the smallest and largest known shark species. Morphological characteristics of this order include two small to moderately large dorsal fins, which may be preceded by a fin spine in some species, the absence of an anal fin, eyes without a nictitating membrane, and five paired gill slits located anteriorly to the pectoral fin. The members of this group inhabit both shallow and deep-water environs, but are generally replaced by requiem (Carcharhinidae) and hammerhead (Sphyrnidae) sharks in warm-temperate and tropical seas. The number of species in this group is expected to increase with continued exploration of the deep sea and improved taxonomic resolution incorporating molecular tools. The members of this group occupy a wide variety of habitats and are found from inshore to over 6000 m in depth. Individual species may be demersal or epipelagic with some species exhibiting pronounced ontogenetic changes in habitat during their lifetime. Four families, six genera, and at least six species occur in the NEP.

The Squalidae are a wide-ranging, benthopelagic family with two genera and 33 species recognized globally. The family shows a high degree of endemism, with only one species (*Squalus acanthias*) known to occur in multiple geographic regions. One genus and a single species occur in the NEP. The North Pacific Spiny Dogfish *Squalus suckleyi* was until recently referred to as the Spiny Dogfish *S. acanthias*, a wide-ranging species, but this latter species is now known to be absent from the North Pacific Ocean (Ebert et al., 2010). The Etmopteridae are the largest family of squaloid sharks with 4 genera and 47 or more species occurring worldwide. Members are benthopelagic, mostly deep-water sharks that occur between 200 and 1500 m deep, though some species may range down to 4500 m (Ebert et al., 2013b). One genus and a single species, the Pacific Black Dogfish *Centroscyllium nigrum* occurs in the NEP. The Somniosidae are one of the widest-ranging groups, occurring in most seas from the tropics to polar seas. Members of this group are generally benthopelagic, although some are pelagic. The group consists of six genera and 17 species, and contains some of the smallest and largest known shark species. The group has representatives of two genera, each with a single species (Pacific Sleeper Shark *Somniosus pacificus* and Velvet Dogfish *Zameus squamulosus*), occurring in the NEP. The Dalatiidae are a small group of seven genera and nine species of dwarf to medium-sized, wide-ranging sharks. Members of this group are mostly pelagic or benthopelagic and occur from near-surface waters down to over 1500 m deep. Two genera, each with a single species (the Dwarf Pelagic Shark *Euprotomicrus bispinatus* and Cookiecutter Shark *Isistius brasiliensis*), are represented in the NEP.

The Squatiniformes (Angel Sharks) comprises a single family (Squatinidae) and genus (*Squatina*) of small, rather morphologically undiverse dorsoventrally flattened sharks. The members of this group are all very similar morphologically and are often misidentified. Members of this group are benthic inhabitants and are often seen partially buried in the sediment. There are approximately 21 described species worldwide, mostly in temperate to subtropical seas, with one species, the Pacific Angelshark *Squatina californica*, represented in the NEP.

The Heterodontiformes (Bullhead or Horn Sharks) are a small group of sharks with one family (Heterodontidae) and a single genus (*Heterodontus*) of nine similar looking species, one species, the Horn Shark *Heterodontus francisci* occurs in the NEP. These are the only living sharks with a dorsal spine preceding each dorsal fin and an anal fin. Members of this group are primarily benthic, nocturnal sharks that tend to rest in caves and rocky crevices during the day, but become quite active at night. They are mostly nearshore sharks occurring from the intertidal down to about 100 m, but at least one species (the Port Jackson Shark *Heterodontus portusjacksoni*) is known to occur down to 275 m (Cox and Francis, 1997).

The order Orectolobiformes (Carpet Sharks) is composed of seven families, 13 genera, and 46 species globally. The only NEP representative of the order is the family Rhincodontidae and genus *Rhincodon*. Members of this group can be distinguished by the presence of nasal barbels, two spineless dorsal fins, and an anal fin. Although primarily benthic tropical to warm-temperate sharks found from intertidal waters to about 200 m deep and usually on rocky or coral reefs, one notable species, the Whale Shark *R. typus* occupies an oceanic habitat. The Whale Shark, based on scattered records, occurs mostly from central California and southwards, but with at least one record from northern California (Ebert et al., 2004).

The Lamniformes (Mackerel Sharks) are a small but diverse group containing seven very distinctive families, 10 genera, and 15 species. Although each of these families appears to be morphologically distinct, they are all united by a number of specific features, including a pointed snout, lack of dorsal fin spines, a similar style of dentition, and oophagous reproduction (in which developing embryos consume eggs as they are ovulated into the uterus). These sharks occupy a wide range of nearshore, coastal, deep-sea, and pelagic habitats from temperate to tropical seas. Six families, eight genera, and 11 species have been recorded from the NEP. The Lamnidae are mostly large, actively swimming, voracious sharks and are among the most media-publicized species. The Alopiidae (Thresher Sharks), with their distinctively long tail, and Laminidae, with the Great White Shark *Carcharodon carcharias*, Shortfin Mako Shark *Isurus oxyrinchus*, and Salmon Shark *Lamna ditropis* being the most common species throughout the region, while the monotypic families Cetorhinidae, Megachasmidae, and Mitsukurinidae, each represented by a single genus and species (see Table 4), are uncommon to rare.

The Carcharhiniformes (Ground Sharks) are the most diverse and largest group of shark-like cartilaginous fishes, comprising eight families, 51 genera, and at least 337 described species. These sharks, despite their wide variability in external morphological appearance, are united by two spineless dorsal fins, an anal fin, a long mouth that extends behind the eyes, nictitating lower eyelids, and (usually) no nasoral grooves. Four families (Scyliorhinidae, Triakidae, Carcharhinidae, and Sphyrnidae), 11 genera, and 18 species occur in the NEP (Table 4).

The Scyliorhinidae (Catsharks) are the largest family of sharks, comprising 17 genera and at least 153 species worldwide; three genera and four species have been reported from the NEP. The group may be subdivided into the colourful, mostly shallow-water species and the uniformly coloured deep-sea forms, which mostly lack any distinctive colour patterns. The number of catshark species is likely to increase with increased exploration, especially in the deep sea. The Triakidae (houndsharks) is a large family consisting of 9 genera and 47 species worldwide, of which three genera and five species are found in the NEP. These are mostly coastal species, several of which are regional endemics. The Carcharhinidae (requiem sharks) are one of the largest shark families, comprising 12 genera and about 59 species worldwide. The family is most diverse in coastal warm-temperate to tropical seas. In the NEP, four genera and six species are known to occur, but except for the Blue Shark *Prionace glauca*, most are uncommon or rare in this region. The Sphyrnidae (hammerhead sharks) are a small group with two genera and nine species, of mostly tropical to warm-temperate species. One genus and three species are found in the NEP, although only the Smooth Hammerhead Shark *Sphyrna zygaena* appears to be common; the other two *Sphyrna* species

Table 4 Selected Natural and Life History Traits of NEP Chondrichthyans

Species	TL_{max} (cm)	Reproductive Mode	Female l_{mat} (cm TL)	Female a_{mat} (Years)	Litter Size (Average)	Longevity (Years)	Population Growth Value	Population Growth Metric	Von Bertalanffy Growth Rate
Hexanchiformes									
Chlamydoselachidae									
Chlamydoselachus anguineus	196	Lecithotrophy	130–135		2–15				
Hexanchidae									
Hexanchus griseus	482	Lecithotrophy	420		22–108				
Notorynchus cepedianus	300	Lecithotrophy		11–21		30	0.23644	r_{max}	0.21–0.29
Echinorhiniformes									
Echinorhinidae									
Echinorhinus cookei	400	Lecithotrophy	250–300		Up to 114				
Squatiniformes									
Squatinidae									
Squatina californica	152	Lecithotrophy	90–100	10–13	1–13 (6)	22–35	0.213528	r_{max}	0.146
Squaliformes									
Dalatiidae									
Euprotomicrus bispinatus	27	Lecithotrophy	22–23						
Isistius brasiliensis	50	Lecithotrophy							
Etmopteridae									
Centroscyllium nigrum	50	Lecithotrophy	43		Up to 7				

Somniosidae									
Somniosus pacificus	700	Lecithotrophy	370–430						
Zameus squamulosus	84	Lecithotrophy	59						
Squalidae									
Squalus suckleyi	130	Lecithotrophy	80–100	36	(9)	>80	0.02–0.03	Intrinsic rebound potential	0.048–0.070
Heterodontiformes									
Heterodontidae									
Heterodontus francisci francisci	96	Oviparous	58			25[a]			
Orectolobiformes									
Rhincodontidae									
Rhincodon typus	2000	Lecithotrophy	900	19–22	Up to 304		0.187519	r_{max}	0.032
Lamniformes									
Alopiidae									
Alopias pelagicus	330	Oophagy	282–292	8–9	2	16–30	0.14516	r_{max}	0.085–0.118
Alopias superciliosus	461	Oophagy	332–355	12–13	2–3	20	0.0929036	r_{max}	0.06–0.18
Alopias vulpinus	573	Oophagy	260–465	3–9	3–4	24–50	1.125	Lambda[b]	0.158–0.215
Cetorhinidae									
Cetorhinus maximus	970	Oophagy	810–980	16–20		50	0.128181	r_{max}	

Continued

Table 4 Selected Natural and Life History Traits of NEP Chondrichthyans—cont'd

Species	TL_{max} (cm)	Reproductive Mode	Female l_{mat} (cm TL)	Female a_{mat} (Years)	Litter Size (Average)	Longevity (Years)	Population Growth Value	Population Growth Metric	Von Bertalanffy Growth Rate
Lamnidae									
Carcharodon carcharias	640	Oophagy	450–500	10–12	2–10	30	0.139571	r_{max}	0.159–0.196
Isurus oxyrinchus	400	Oophagy	275–293	19–21	(10–18)	29–32	0.116395	r_{max}	0.17–0.19
Isurus paucus	426.7	Oophagy	>245		2–8				
Lamna ditropis	261	Oophagy	180[c]	6–9	3–5	17–20	0.218735	r_{max}	0.17–0.23
Megachasmidae									
Megachasma pelagios	577	Oophagy							
Mitsukurinidae									
Mitsukurina owstoni	617	Oophagy							
Odontaspididae									
Odontaspis ferox	450	Oophagy	300–350						
Carcharhiniformes									
Carcharhinidae									
Carcharhinus brachyurus	295	Lecithotrophy	227–244	19–20	7–24 (15)		0.123547	r_{max}	0.0385
Carcharhinus longimanus	400	Lecithotrophy	170–190	3–5	Up to 15		0.212198	r_{max}	0.103
Carcharhinus obscurus	360	Matrotrophy	235[d]	17.6	3–16	40–50	0.0915137	r_{max}	0.039

Galeocerdo cuvier	403	Lecithotrophy	250–350		(30–35)	45–50	0.271078	r_{max}	0.31
Prionace glauca	380	Matrotrophy	173–221	5–7	−35	20	0.577147	r_{max}	0.18–0.25
Rhizoprionodon longurio	154	Matrotrophy	103		1–12 (7.4)				
Scyliorhinidae									
Apristurus brunneus	66	Oviparity	50.1						
Apristurus kampae	64.7	Oviparity	49						
Cephaloscyllium ventriosum	110	Oviparity							
Parmaturus xaniurus	61	Oviparity	42.5–50						
Sphyrnidae									
Sphyrna lewini	346	Lecithotrophy	210–250	5.8–15	−12.9	30.5	0.464941	r_{max}	0.073–0.013
Sphyrna tiburo	150	Matrotrophy	80–95	2–3	(9–10)	17.9	0.556761	r_{max}	0.280–0.69
Sphyrna zygaena	440	Matrotrophy	265		20–49 (32)	>20			0.06–0.07
Triakidae									
Galeorhinus galeus	165	Lecithotrophy	134–140	10–15	−35	60	0.195574	r_{max}	0.086–0.104
Mustelus californicus	163	Lecithotrophy	70–90	2–3	3–16	9	0.673236	r_{max}	0.168
Mustelus henlei	100	Lecithotrophy	51–66	2–3	1–21	13	0.500002	r_{max}	0.224
Mustelus lunulatus	175	Matrotrophy	103.2	6	6–19	22			
Triakis semifasciata	180	Lecithotrophy	105–135	10–15	7–36	25–30	0.271586	r_{max}	0.082

Continued

Table 4 Selected Natural and Life History Traits of NEP Chondrichthyans—cont'd

Species	TL_{max} (cm)	Reproductive Mode	Female l_{mat} (cm TL)	Female a_{mat} (Years)	Litter Size (Average)	Longevity (Years)	Population Growth Value	Population Growth Metric	Von Bertalanffy Growth Rate
Rajiformes									
Arhynchobatidae									
Bathyraja abyssicola	157	Oviparity							
Bathyraja aleutica	154	Oviparity	112–133	9–14	3–60	19	1.252	Lambda[b]	0.11–0.13
Bathyraja interrupta	82.5	Oviparity	70–72	7.5		21	1.36	Lambda[b]	0.08
Bathyraja kincaidii	56	Oviparity	46–50	4		17–18			0.185–0.237
Bathyraja lindbergi	102	Oviparity		21		35	1.11	Lambda[b]	0.04
Bathyraja maculata	147	Oviparity		26		36	1.079	Lambda[b]	0.02
Bathyraja mariposa	76	Oviparity	76						
Bathyraja microtrachys	91	Oviparity	60–70						
Bathyraja minispinosa	82.5	Oviparity	66.8			37	1.096	Lambda[b]	0.02
Bathyraja panthera	139	Oviparity							
Bathyraja parmifera	111	Oviparity	80.9–95.4	8–10		20			0.1
Bathyraja spinosissima	203	Oviparity							
Bathyraja taranetzi	77	Oviparity	61.4	9.1		14	1.116	Lambda[b]	
Bathyraja trachura	91	Oviparity	72–80	14–25		36	1.048	Lambda[b]	0.08
Bathyraja violacea	107	Oviparity	61.2–76.4	5–6					0.049

Rajidae									
Amblyraja badia	99	Oviparity							
Beringraja binoculata	244	Oviparity	90–150	8–10	1–7	26	0.814461	r_{max}	0.07–0.1145
Raja inornata	76	Oviparity	50–52	2.9		9			
Raja rhina	203.9	Oviparity	70–100	10–12		26	1.202	Lambda[b]	0.046–0.2
Raja stellulata	76.1	Oviparity	63.2	11.2		15			0.013
Rhinobatidae									
Pseudobatos productus	156	Lecithotrophy	57–99	7	1–16	11	0.268414	r_{max}	
Zapteryx exasperata	97	Lecithotrophy	57–77		2–13				
Torpediniformes									
Torpedinidae									
Tetronarce californica	140	Lecithotrophy	73.1	9		16	0.288543	r_{max}	0.073–0.173
Myliobatiformes									
Dasyatidae									
Hypanus dipterurus	100[e]	Lecithotrophy	43.4	10	1–4 (2.7)	28	0.133123	r_{max}	0.1–0.55
Pteroplatytrygon violacea	80[e]	Matrotrophy	39–50	3	4–13	10	0.480792	r_{max}	
Gymnuridae									
Gymnura marmorata	100 (125[e])	Matrotrophy	62		4–16				
Mobulidae									
Manta birostris	700	Lecithotrophy	410[e]	8–10	1	40	0.116	r_{max}	
Mobula japanica	310	Lecithotrophy	207[e]	5–6	1		0.116	r_{max}	0.12

Continued

Table 4 Selected Natural and Life History Traits of NEP Chondrichthyans—cont'd

Species	TL_{max} (cm)	Reproductive Mode	Female l_{mat} (cm TL)	Female a_{mat} (Years)	Litter Size (Average)	Longevity (Years)	Population Growth Value	Population Growth Metric	Von Bertalanffy Growth Rate
Myliobatidae									
Myliobatis californicus	180	Matrotrophy	88.1[e]	5	Up to 12		0.296264	r_{max}	0.0995–0.229
Platyrhinidae									
Platyrhinoidis triseriata	91	Lecithotrophy	48		1–15				
Urolophidae									
Urobatis halleri	33.3	Matrotrophy	15[e]	3.5	1–11	14			0.09–0.15
Chimaeriformes									
Chimaeridae									
Hydrolagus cf. *trolli*	111	Oviparity	55[f]						
Hydrolagus colliei	63	Oviparity	19[g]						
Hydrolagus melanophasma	120	Oviparity							
Rhinochimaeridae									
Harriotta raleighana	102.5	Oviparity	30[f]						

[a]Unconfirmed.
[b]Finite annual population growth rate.
[c]cm precaudal length, not total length.
[d]cm fork length, not total length.
[e]cm disc width, not total length.
[f]cm body length.
[g]cm snout to vent length.
Data are from IUCN Red List Assessments supplemented by primary literature.

are occasionally observed off southern California during warm water El Niño events.

The Torpediniformes (Electric, Thornback, and Fan Rays) are a moderately sized group of specialized batoids, generally characterized by a thick, flabby, oval to rounded disc, a short, broadly rounded to truncated snout, soft, loose, or firm skin with minute to no dermal denticles, a relatively short, thick tail, zero to two dorsal fins, and a well-developed caudal fin. However, the family Platyrhinidae was recently reassigned to this order (Eschmeyer et al., 2016), based mainly on molecular and not morphological traits. Externally, the Platyrhinidae morphologically are firm bodied, with a heart-shaped disc, and have one to three rows of large hook-like thorns along midback (Ebert, 2003). Further research into the relationship of the electric rays to platyrhinids is required to clarify the status of these seemingly morphologically diverse groups. Worldwide, the order Torpediniformes, with the inclusion of the Platyrhinidae, presently has five families, 15 genera, and 74 species recognized, of which 2 families, 2 genera, and 2 species occur in the NEP.

The Rhinopristiformes (Sawfish, Guitarfish, and Wedgefish) are comprised of five families, 12 genera, and about 66 species. Morphologically, the group is characterized by a wedge-shaped disc, a small, single row of thorns along midback, and a first dorsal fin closer to pelvic fins than to caudal fin. The NEP has two families, two genera, and two species represented; the Shovelnose Guitarfish *Pseudobatos productus* (formerly *Rhinobatos productus*) was recently reassigned to a new genus and the Banded Guitarfish *Zapteryx exasperata* was reallocated to the newly erected family Trygonorrhinidae (Last et al., 2016).

The Rajiformes (skates) are the largest Chondrichthyan order globally, comprising four families, 38 genera, and nearly 300 species, a number likely to increase as new species are described. Members of this group are characterized by a rhomboid-shaped disc, a long, broadly angular or rounded preoral snout, five paired gill openings, no anal fin, a long slender tail tapering posteriorly, usually with one or more rows of enlarged thorns or smaller thornlets, two small dorsal fins set well behind the pelvic fins, and a small caudal fin that may be reduced or absent in some species. In the NEP two families, four genera, and at least 20 species represent skates, although the number of genera and species in the region is likely to fluxuate, as some species appear to represent complexes of one or more species.

The two NEP skate families can be easily distinguished in the field based on whether the snout is soft and flexible (Arhynchobatidae) or rigid and firm

(Rajidae). Furthermore, the softnose skates are mostly deep-sea species, occurring below the 200 m contour, and have their greatest diversity and abundance at higher latitudes in cold-temperate and polar seas (Ebert and Winton, 2010). Worldwide, the softnose skates have 17 genera and 105 species, and in the NEP are represented by a single genus and 15 species. All species occurring in the NEP are considered deep-sea inhabitants, with the exception of the Alaska Skate *Bathyraja parmifera*, which is the most abundant chondrichthyan on the EBS shelf (Stevenson et al., 2008). Three species, Deepsea Skate *Bathyraja abyssicola*, Fine-spined Skate *Bathyraja microtrachys*, and Pacific White Skate *Bathyraja spinosissima*, by contrast, are known to occur to at least 2900–3300 m deep (Ebert, 2003; unpublished data). The hardnose skates are the most diverse skate family with 21 genera and 158 species, and with 3 genera and 5 species occurring in the NEP. The hardnose skates in the NEP primarily occur shallower than the softnose skates, with most species commonly occurring from nearshore to the outer continental shelf and upper continental slope, but at less than 400 m deep. The Broad Skate *Amblyraja badia* is an exception though as this hardnose skate species occurs from 846 to over 2300 m deep (Ebert, 2003).

The Myliobatiformes (Eaglerays, Devilrays, Stingrays) are the second largest group of batoids and are characterized by their enlarged, expanded pectoral fins being completely fused to the head and trunk (except for the Myliobatidae), a disc shape ranging from diamond-shaped to subrhombic, or lozenge-shaped, the presence or absence of a single dorsal fin, a long whip-like tail, and one or more stinging spines on the upper tail base. Worldwide there are 12 families, 40 genera, and over 260 species recognized. Of these totals, five families, seven genera, and seven species occur in the NEP. The order mostly replaces the skates in warm-temperate to tropical seas (White and Sommerville, 2010). The majority of NEP species, five of seven, occur off southern California with their abundance and populations increasing to the south in the warm-temperate waters of Baja California.

The Chimaeriformes (Chimaeras, Ghost Sharks, Ratfish, or Silver Sharks) are a small group of chondrichthyans comprised of three families, six genera, and 51 described species. The chimaeras are readily distinguished by their elongated tapering body, filamentous whip-like tails, smooth scales, large venomous dorsal fin spine, broad wing-like pectoral fins, and open lateral line canals, which appear as grooves on the head and along the trunk. Chimaeras are sluggish swimming species usually found along outer continental shelves and upper slopes, although some species, especially members of the Callorhinchidae, are common in nearshore coastal waters. Two families, two genera, and four species occur in the NEP. The Chimaeridae, which is the most species-rich family, has a single genus (*Hydrolagus*) and is represented by three of 43 species occurring in the NEP; one of the species (*Hydrolagus* cf. *trolli*), however, has only been tentatively identified based on deep-sea video footage (Reichert et al., 2016). The family Rhinochimaeridae is a wide-ranging group with one representative genus *Harriotta* and a single species the Longnose Chimaera *Harriotta raleighana* in the NEP.

7. LIFE HISTORIES

Life histories describe the speed of life and include traits related to survival, growth, and reproduction (Roff, 2002; Stearns, 1992). Life history theory suggests that due to constraints and trade-offs among life history traits, species fall along a continuum of slow to fast life histories (Charnov, 1993; Reynolds et al., 2001; Stearns, 1983, 1992). Slow life histories describe those species that have slower growth, lower reproductive output, long gestation times, later ages at maturity, higher longevities (and thus longer generation times), larger body sizes, and lower population growth rates. Fast life histories describe those species that are on the opposite end of the continuum with faster growth, higher reproductive output, shorter gestation times, earlier ages at maturity, shorter lifespans, shorter generation times, smaller body sizes, and higher population growth rates (Denney et al., 2002; Gunderson, 1980; Pianka, 1970). Although chondrichthyans generally are classified as a group with slower life histories than other vertebrates (e.g. Hoenig and Gruber, 1990; Hutchings et al., 2012), much variation exists among species (Table 4). For example, this variation is evident when contrasting the 18- to 24-month gestation length of the North Pacific Spiny Dogfish (*Squalus suckleyi*) with the 3-month gestation length of the Round Stingray (*Urobatis halleri*), both of which are present in the NEP (Ketchen, 1986; Mull et al., 2010; Tribuzio and Kruse, 2012).

8. REPRODUCTION

Chondrichthyans exhibit a diversity of reproductive modes compared to other vertebrate groups (Wourms, 1977; Wourms and Demski, 1993). These reproductive modes can be largely categorized into two major groups, viviparity and oviparity, although variations exist within each (Musick and

Ellis, 2005). Viviparous species retain the developing embryo inside the uterus, and those that exhibit this mode of reproduction generally have two development strategies, lecithotrophy or matrotrophy. Species that display lecithotrophic viviparity nourish embryos with a yolk sac during development, but have no maternal investment postfertilization, i.e., no nutrition source from the female's body. Matrotrophic viviparous species nourish developing embryos by either excreting a protein- or lipid-rich histotroph or producing unfertilized eggs to be consumed by the developing embryo; the latter is known as oophagy. In contrast, oviparous species deposit or "lay" egg cases on a benthic substrate and the morphology of the egg case is unique to each species. These egg cases typically have a tough exterior for protecting developing embryos and often have morphological adaptations, such as external filamentous tendrils or hooked protrusions, for adhering to the substrate (Musick and Ellis, 2005).

Of the 77 species in the NEP, 62% are viviparous and 38% are oviparous (Table 4), which is not that dissimilar to the global proportion of oviparous chondrichthyans (57% viviparous and 43% oviparous) (Ebert and Winton, 2010). Of viviparous species in the NEP, 56% are lecithotrophic viviparous and 44% are matrotrophic viviparous. All species of the orders Hexanchiformes, Echinorhiniformes, Squatiniformes, and Squaliformes in the NEP exhibit lecithotrophy. Additionally, almost half (44%) of the 18 carcharhiniform species and half (50%) of the 8 myliobatid species in the NEP are lecithotrophic (Table 4). The remaining four myliobatid species (50%) are matrotrophic, as are the six carcharhiniform and 11 lamniform species, the latter exhibiting oophagy, a type of matrotrophy. The oviparous species in the NEP include the one heterodontiform species, the four chimaeriform species, the 20 rajiform species, and the four species in the Family Scyliorhinidae within Carcharhiniformes.

Little is known about the frequency and duration of the reproductive cycle of chondrichthyans globally, and species inhabiting the NEP are no exception. Oviparous chondrichthyans generally have one of three reproductive cycles, and these are somewhat similar to those in viviparous species (Ebert and Winton, 2010; Wourms, 1977). The three cycles are: (1) continuous (active throughout the year), (2) annual or biennial, or (3) a partially defined annual cycle with one or two peaks amidst mostly continuous reproduction (Wourms, 1977). For species in the NEP whose reproductive cycle is known (38 of 77 species), most have an annual or biennial reproductive cycle, or reproduce continuously. The oviparous Rajiformes mostly reproduce continuously and the viviparous (both lecithotrophy and matrotrophy)

Carcharhiniformes and Myliobatiformes, for which data are available, largely reproduce annually or biennially.

Fecundity can be estimated by multiple ways including litter size, annual reproductive output (offspring per year), or some measure of offspring production (e.g. grams offspring per year). Estimates of these fecundity metrics are sparse for chondrichthyans as a group, especially for oviparous species. For species present in the NEP, around half have some published fecundity estimates, and these are mostly viviparous species. Litter size is the most commonly reported fecundity metric for species in the NEP, and ranges from one pup per litter in the mobulid rays Manta ray (*Manta birostris*) and Spinetail Devilray (*Mobula japanica*) to over 300 pups per litter in the Whale Shark (*R. typus*) (Table 4), which is the largest litter size reported for any shark species (Beebe and Tee-Van, 1941; Joung et al., 1996). For those species in the NEP whose annual reproductive output can be calculated (mostly those with a biennial or annual reproductive cycle), this fecundity metric ranges from 2.7 offspring per year in the Diamond Stingray (*Hypanus dipterurus*) to 12.9 in the Scalloped Hammerhead (*Sphyrna lewini*) (Liu and Chen, 1999; Smith, 2004). Fecundity estimates for oviparous species are largely from captive studies, and these studies suggest that some species have high fecundity on both annual and lifetime scales (Ebert, 2005; Holden et al., 1971; Ishihara et al., 2002). Some species in the NEP have been found to reproductively senesce, or have a decline in fecundity with increasing length and/or age (Ebert, 2005). For example, larger and likely older individuals of Aleutian Skate (*Bathyraja aleutica*), Commander Skate (*Bathyraja lindbergi*), and Whitebrow Skate (*Bathyraja minispinosa*) were found to have fewer oocytes or atrophied ovaries (Ebert, 2005). This has also been documented in elasmobranchs of various orders around the world to date (Ebert, 2005; Peres and Vooren, 1991).

9. AGE AND GROWTH

Data from age and somatic growth studies such as growth rates, age or size at maturity, maximum theoretical length, and longevity are among the most fundamental life history traits of organisms, and are essential in estimating other demographic parameters such as population growth rates, mortality estimates, and fecundity rates (Cortés, 1997; Frisk et al., 2005; Jennings et al., 1998; Pauly, 1980). This information provides insight into the status of a population and its ability to recover from exploitation (Cortés, 2002; Hutchings et al., 2012; Juan-Jordá et al., 2013). Thus, having these data are crucial when developing effective management strategies and conservation efforts, especially for chondrichthyans, which are one of the most threatened groups of marine life (Dulvy et al., 2008, 2014a; McClenachan et al., 2012).

Chondrichthyans are aged similarly to other fishes, in that growth band patterns are examined in calcified structures (Cailliet, 1990; Cailliet and Goldman, 2004; Goldman, 2005; see chapter "Age and growth of elasmobranchs and applications to fisheries management and conservation in the Northeast Pacific Ocean" by Matta et al., this volume). Vertebral centra are the most commonly used structure for ageing cartilaginous fishes; caudal thorns and fin spines have also been used with success in some species (Cailliet and Goldman, 2004; Davis et al., 2007; Ketchen, 1975). Growth bands in neural arches and the upper jaw have also been explored as other structures for potential use in ageing elasmobranchs (McFarlane et al., 2002; Tanaka, 1990). Methods for assessing age and somatic growth include sectioning these structures (fin spines are often just sanded) to visualize and examine banding patterns, and then verifying and validating the resulting age estimates (Cailliet and Goldman, 2004). Assessing, verifying, and validating the age and somatic growth of chondrichthyans can be problematic for many reasons (Cailliet and Goldman, 2004; Natanson and Cailliet, 1990). For example, some species have too poorly calcified vertebral centra or do not deposit predictable growth bands in their vertebrae (McFarlane et al., 2002; Natanson and Cailliet, 1990). Validating age and somatic growth estimates are even more problematic as this typically involves chemically tagging wild fish, mark-recapture studies of individuals with known ages, and/or bomb carbon dating (Ardizzone et al., 2006; Cailliet, 1990; Cailliet and Goldman, 2004). Because of the difficulty associated with validating age and somatic growth estimates, verification methods such as centrum edge analysis and relative marginal increment analysis are used (see Cailliet and Goldman, 2004; chapter "Age and growth of elasmobranchs and applications to fisheries management and conservation in the Northeast Pacific Ocean" by Matta et al., this volume).

Although somatic growth data are largely missing for many chondrichthyan species, those in the NEP are relatively well studied with respect to these life history traits. Over half (53%) of all species in the NEP have some published somatic growth rate data, but it is important to note that for many of these species that are not endemic to the NEP, not all were estimated on specimens collected in the NEP region. Importantly, somatic growth rates, as well as most other life history traits, have been known to vary widely among populations of the same species, largely due to different environmental conditions (Licandeo and Cerna, 2007; Lombardi-Carlson et al., 2003; Neer and Thompson, 2005). For example, somatic growth rates in the Longnose skate (*Raja rhina*) vary with geographic region and are lower at the northern end of this skate's distribution and increase towards the southern end (Gburski et al., 2007; McFarlane and King, 2006; Zeiner, 1991).

Many different models exist for estimating somatic growth parameters in fishes, and the von Bertalanffy and Gompertz growth models are by far the most commonly applied (Cailliet and Goldman, 2004; Goldman, 2005). For NEP chondrichthyans, most age and somatic growth studies use the von Bertalanffy growth function (von Bertalanffy, 1938) and report k, the growth coefficient, which describes the average rate at which an organism in the population reaches its maximum length or size from its length or size at birth (Cailliet and Goldman, 2004; Goldman, 2005). Thirty-nine of 77 (51%) NEP chondrichthyans have published k values, which vary widely across species (Table 4). The Starry Skate (*Raja stellulata*) has the lowest growth coefficient ($k=0.013$), and two other skates, Whitebrow Skate and White-blotched Skate (*Bathyraja maculata*), are just slightly higher ($k=0.02$) (Ainsley et al., 2011; James et al., 2014; Maurer, 2009). The highest reported growth coefficient (0.280–0.69) for a species found in the NEP is the Bonnethead Shark (*Sphyrna tiburo*), but this estimate came from specimens collected in northwest Florida, USA (Carlson and Parsons, 1997). Most NEP species that have published growth coefficients are either in the Carcharhiniformes (12 species) or the Rajiformes (12 species). Of the top five slowest growing species in the NEP, three are batoids (Table 4). Of the top five fastest growing species, two are rajiform species, two are carcharhiniform species, and one species is from Lamniformes.

Similar to the variation in somatic growth, substantial variation in age and length at maturity and longevity exists among chondrichthyans. Female age at maturity has been reported in the literature for 56% of species in the NEP, and male age at maturity for 53% of species (Table 4). In terms of female age at maturity, the faster-maturing species in the NEP are from Carcharhiniformes (Grey Smooth-hound (*Mustelus californicus*), Brown Smooth-hound (*Mustelus henlei*), and Bonnethead shark) and these species mature around 2–3 years of age (Carlson and Parsons, 1997; Yudin and Cailliet, 1990). The latest-maturing species in terms of female age at maturity is the North Pacific Spiny Dogfish, which does not become sexually mature until around age 36 (ranges from 23 to 36 in the literature)

(Ketchen, 1975; McFarlane et al., 2010; Tribuzio and Kruse, 2012). Other later-maturing species in terms of female age at maturity include the White-blotched skate and Commander skate (26 and 21 years, respectively; Maurer, 2009), Roughtail Skate (*Bathyraja trachura*) (14–25 years; Davis et al., 2007), and the sharks Dusky Shark (*Carcharhinus obscurus*) (17.6 years; Natanson et al., 1995), Shortfin Mako shark (19–21 years; Campana et al., 2002; Ribot-Carballal et al., 2005), and Copper Shark (*Carcharhinus brachyurus*) (19–20 years; Walter and Ebert, 1991). Many of these species also have reported length at maturity, and most mature at 60%–80% of their total maximum reported or theoretical size. For the 42 of 77 (55%) species in the NEP for which longevity has been reported, it ranges from 9 years old in Grey smooth-hound (Yudin and Cailliet, 1990) to over 80 years in North Pacific Spiny Dogfish (McFarlane and Beamish, 1987), with most reaching over 20 years of age.

10. POPULATION DYNAMICS, DEMOGRAPHY, AND EXTINCTION RISK

Population dynamics and demography encompass the spatial and temporal changes in the abundance of a population due to ecological, environmental, and genetic variability (Cortés, 2004; Cortés et al., 2012; Dulvy and Forrest, 2010). Strictly speaking, demographic parameters are the vital rates of birth, death, immigration, and emigration, and population dynamics describe how the population, typically at the resolution of age or stage-specific cohorts, changes in response to these demographic parameters (Cortés, 2004; Cortés et al., 2012; Dulvy and Forrest, 2010). Understanding the population dynamics and demography of a species or population is critical as these data not only form the basis of stock assessments but also inform a species' vulnerability to decline as a result of natural and anthropogenic stressors (Cortés, 1998; Dulvy et al., 2004; Reynolds et al., 2001).

Modelling population dynamics and estimating demographic metrics can be difficult for chondrichthyans due to the data-poor and threatened nature of many members of this group (Cortés, 2004; Cortés et al., 2012; Dulvy et al., 2014a). Estimating these metrics for this group of fishes is typically conducted using life tables and population matrix models (see Cortés, 2004; Cortés et al., 2012; Dulvy and Forrest, 2010; Simpfendorfer, 2005). Some metrics (e.g. population growth rates) can be calculated using life history traits (see Hutchings et al., 2012; Pardo et al., 2016). In the last few years, great progress has been made in improving the quantity and quality of chondrichthyan demographic data (Barnett et al., 2013; Cortés et al., 2012; Pardo et al., 2016). Unfortunately, understanding extinction risk is still problematic and often correlate traits, such as life history traits, are used (Denney et al., 2002; Dulvy et al., 2014a; Jennings et al., 1999). Despite the use of body size and other size-related traits to inform extinction risk, recent work suggests that time-related traits, such as population growth rates, are better correlates (Barnett et al., 2013; Jennings et al., 1998; Juan-Jordá et al., 2015). In fact, many studies use some measure of population growth rate as a measure of species' vulnerability to decline or extinction risk (García et al., 2008; Hutchings et al., 2012; Juan-Jordá et al., 2013).

For those species in the NEP, 40 (52%) have some measure of population growth rate reported in the literature (Table 4). The orders with the highest number of species with published population growth rates are the Carcharhiniformes ($n=11$), Rajiformes ($n=10$), Lamniformes ($n=7$), and Myliobatiformes ($n=5$). The most commonly reported population growth rate metric for NEP species ($n=28$, 70%) is the maximum intrinsic rate of population increase (r_{max}), which is the productivity of a depleted population in which density-dependent regulation is absent (Myers et al., 1997, 1999; Pardo et al., 2016). Other reported population growth rate metrics for species in this region are the intrinsic rebound potential (Tribuzio et al., 2011) and the finite rate of annual population increase (λ; Barnett et al., 2013); see Dulvy et al. (2004) and Simpfendorfer (2005) for information about these and other population growth rate metrics. The range of r_{max} values for species in the NEP spans 0.092 in the Dusky shark to 0.81 in the Big Skate. Four of the top five species with the highest r_{max} values are in the order Carcharhiniformes (Brown smooth-hound shark, 0.5; Bonnethead shark, 0.557; Blue shark, 0.577; and Grey smooth-hound shark, 0.673) (Table 4). The five species with the lowest r_{max} values are comprised of two species in the Lamniformes (Bigeye Thresher shark, 0.093; Shortfin Mako shark, 0.116), two species in the Myliobatiformes (Spinetail Devilray and Manta ray, 0.116 for both [this r_{max} value was calculated for a "generic manta ray", Dulvy et al., 2014b]), and one carcharhiniform (Dusky shark, 0.092) (Table 4). As with the somatic growth rates and other life history traits reported earlier, not all population growth rates were estimated from specimens within the NEP region. Life history traits are known to vary widely, even intraspecifically depending upon population and geographic region, and this should be taken into account when developing management strategies or using these life history traits as correlates of extinction risk.

11. DIET, TROPHIC ECOLOGY, AND TROPHIC LEVEL

Diet and trophic ecology provide insight into predator–prey interactions: community linkages, food web dynamics, and energy transfer in marine environments (Heithaus et al., 2010; Vaudo and Heithaus, 2011; Wetherbee et al., 2012; see chapter "Diet composition and trophic ecology of Northeast Pacific sharks" by Bizzarro et al., this volume). Diet is typically assessed through analysing stomach contents, but this method only provides short-term resolution (Cortés, 1999; Hyslop, 1980; Peterson and Fry, 1987). Estimating stable isotope signatures of various organismal tissues of both predators and their prey is an important complement to traditional stomach content analysis as it provides a longer-term picture of the trophic ecology of an organism (Peterson and Fry, 1987; see chapter "Stable isotope applications for understanding shark ecology in the Northeast Pacific Ocean" by Reum et al., this volume). Stable isotope analysis is also a useful tool for elucidating more than just trophic ecology and how it shifts ontogenetically, temporally, and spatially; movements and habitat use, as well as how these shift, have been commonly reported in the chondrichthyan literature (Carlisle et al., 2011, 2014, 2015; Estrada et al., 2006; Kim et al., 2012). In addition, both stable isotope and stomach content analysis can be used to estimate the trophic level or position of an organism, which is useful in interspecific and intraspecific comparisons to further understanding food web structure and other ecosystem trophodynamics (Boyle et al., 2012; Cortés, 1999; Ebert and Bizzarro, 2007; see chapter "Diet composition and trophic ecology of Northeast Pacific sharks" by Bizzarro et al., this volume).

Chondrichthyans are carnivorous and feed on fishes, cephalopods, crustaceans, zooplankton, and annelid worms (Cortés, 1999; Ebert and Bizzarro, 2007; Wetherbee et al., 2012). For the 61 species (79%) of chondrichthyans in the NEP for which diet has been reported, the most common prey items are teleost fishes, cephalopods, and decapod crustaceans. Other common items include molluscs, amphipod crustaceans, and annelid worms. Five species (Basking Shark *Cetorhinus maximus*, Manta Ray, Megamouth Shark *Megachasma pelagios*, Spinetail Devilray, and Whale Shark *Rhincodon typus*) filter-feed on zooplankton (Baduini, 1995; Beebe and Tee-Van, 1941; Cortés, 1999; Eckert and Stewart, 2001; Notarbartolo-di-Sciara, 1988).

Seabirds, marine mammals, and other chondrichthyan fishes form the diet of some NEP chondrichthyans (White Shark *Carcharodon carcharias*, Tiger Shark *Galeocerdo cuvier*, and Broadnose Sevengill Shark *Notorynchus cepedianus*) (Cortés, 1999). Many species scavenge (e.g. *N. cepedianus* and Bluntnose Sixgill Shark *Hexanchus griseus*) and one (Cookiecutter Shark *Isistius brasiliensis*) even bites plugs of tissues off of large predators (Cortés, 1999). Stable isotope studies have been conducted on 39 of 77 (51%) of NEP species and have yielded important information that was not apparent in diet studies (see chapter "Stable isotope applications for understanding shark ecology in the Northeast Pacific Ocean" by Reum et al., this volume). For example, *S. suckleyi* feeds predominantly on fishes and cephalopods, but stable isotope analysis showed much lower carbon and nitrogen signatures than would be expected based on stomach contents alone due to their seasonal consumption of euphausiids (Bigman, unpublished data).

Estimates of trophic levels have been reported in the literature for 47 of 77 (61%) of chondrichthyans in the NEP. There have been two major studies that have estimated trophic levels of large groups of chondrichthyans, many of which are present in the NEP (Cortés, 1999; Ebert and Bizzarro, 2007; see updated information in chapter "Diet composition and trophic ecology of Northeast Pacific sharks" by Bizzarro et al., this volume). Cortés (1999) calculated standardized diet compositions and trophic levels for 149 species of sharks, and Ebert and Bizzarro (2007) estimated trophic levels for 60 skate species. These studies have found that chondrichthyans generally occupy trophic levels ranging from three to four, which places them on a similar level to that of teleosts, seabirds, and marine mammals (Cortés, 1999; Ebert and Bizzarro, 2007). These two studies used stomach content data to estimate trophic levels; some previous work has found that trophic level does not differ depending upon method (using stomach content or stable isotope analysis data) (Boyle et al., 2012; Estrada et al., 2003), but not all studies have found similar results (Fisk et al., 2002; Hussey et al., 2011).

Trophic levels for chondrichthyans in the NEP ranged from 3.2 to 4.7 (Cortés, 1999; see chapter "Diet composition and trophic ecology of Northeast Pacific sharks" by Bizzarro et al., this volume). Orders with the most species having published trophic levels were the Carcharhiniformes, Rajiformes, and Lamniformes. As a group, species in the

Rajiformes had the lowest trophic levels of any order, and species in the Carcharhiniformes and Lamniformes had the highest trophic levels. The lowest trophic levels (<3.5) belonged to *S. tiburo* (3.2), Horn shark *H. francisi* (3.2), *C. maximus* (3.2), *M. pelagios* (3.4), and *Mobula japonica* (3.43), three of which are filter feeders (*C. maximus*, *M. pelagios*, and *M. japonica*). The highest trophic levels (>4.3) belonged to *H. griseus* (4.3), *I. oxyrinchus* (4.3), *I. brasiliensis* (4.3), *C. carcharias* (4.5), and *N. cepedianus* (4.7). There was commonly a range of trophic levels within families, but the three thresher shark species (family Alopiidae) all had similar trophic levels (~4.2), as did the Lamnidae (~4.4). The only group with no known published trophic levels is the Chimaeras.

12. HABITAT ASSOCIATION

Most chondrichthyan species in the NEP primarily utilize benthic or benthopelagic habitat (70%), with only 23 of 77 species living in pelagic waters. This is in contrast to the global proportion of 90% benthic or benthopelagic chondrichthyans (Compagno, 1984). Generally, skates, squaloids, and Chimaeras live close to or on the bottom of the seafloor (Compagno, 1984), and quite a few species from these groups are present in the NEP. In fact, all of the rajiform ($n=22$), squaliform ($n=6$), and chimaeriform species ($n=4$) in the NEP are benthic. The Carcharhiniformes are split evenly into benthic and pelagic habitat groupings ($n=9$, respectively). The pelagic proportion is largely comprised of those Carcharhiniformes species, nine Lamniformes, and four Myliobatiformes. Few NEP chondrichthyan species use coastal or nearshore habitats as adults. As habitat preferences change spatially, temporally, and ontogenetically, it is important to understand these changes as they have management and conservation implications (Grubbs, 2010; Morris, 2003). For example, *S. suckleyi* segregates by size and sex, and these groups usually occupy different areas; juveniles occupy pelagic waters, and adults are demersal (Ketchen, 1986). Salmon shark *Lamna ditropis* shift from oceanic to neritic habitats around the age of maturity (Carlisle et al., 2014), as well as undergoing a directed northern migration from pupping grounds off California to adult feeding grounds of British Columbia and Alaska. Despite these examples, little is generally known about shifts in habitat of chondrichthyans and it is important to note that reported habitat preferences do not likely fully encompass habitat use of a particular species.

13. FISHERIES AND CONSERVATION

13.1 Fisheries

The two leading threats to the world's chondrichthyans are overexploitation—through targeted and incidental capture—and habitat loss (Dulvy et al., 2014a). Therefore, some of the most effective tools for species conservation for chondrichthyans are fisheries management and habitat protection. When the status of the world's fisheries was examined at a global scale, the NEP stood out as a becon for successful fisheries management and marine protected area (MPA) implementation (Worm et al., 2009). The EBS has consistently been managed below its multispecies maximum sustainable yield, while exploitation rates in the CAC have declined substantially while biomass has increased (Worm et al., 2009). MPAs have been designated and successfully implemented in the NEP at the State, Provincial, and Federal level. In particular, California's Marine Life Protection Act (MLPA) initiative has been heralded as a model in MPA development and implementation, driven by strong legal mandates, political leadership, sound scientific knowledge, and effective stakeholder participation (Jones et al., 2013). California's MPAs encompass approximately 18% of state waters in the Central Coast region and have shifted trawl fishing effort to deeper waters (California Department of Fish and Wildlife, 2015). In Washington State the first MPA was designated in 1907 and 127 MPAs managed by a variety of jurisdictions current provide protection for over 640,000 acres of marine and nearshore habitat (Van Cleve et al., 2009). In 2010, and reaffirmed in 2015, the United Nations set a suite of Sustainable Development Goals for the world at large, which includes an aggressive goal to conserve 10% of coastal and marine waters as MPAs by 2020 (United Nations, 2015), a goal commonly referred to as Aichi Target 11 (https://www.cbd.int/sp/targets/default.shtml). Canada has embraced this goal and set an additional interim goal of having 5% of Canadian waters, including significant portions of the territorial waters of British Columbia in the NEP (Department of Fisheries and Oceans Canada, 2014), set aside as MPAs by 2017 (http://www.dfo-mpo.gc.ca/oceans/conservation/plan-eng.html). Management and conservation of NEP chondrichthyan species involves management bodies in the USA and Canada, as well as international bodies, and species that are distributed in the southern CAC also rely on management from Mexico. However, for the purposes of this chapter we will focus mainly on the USA and Canada.

Globally, the most highly threatened chondrichthyans are coastal and continental shelf-dwelling rays that are exposed to shallow-water fisheries (Dulvy et al., 2014a). Skates, the dominant chondrichthyan fauna of the EBS and GOA, vary in depth range, with most of these species occupying deep waters (i.e. the Deepsea Skate *Bathyraja abyssicola* and Fine-spined Skate *B. microtrachys*). However, some skate species are most abundant in shallower waters (<200 m) including the Big Skate *Beringraja binoculata*, the Longnose Skate *Raja rhina*, and the Alaska Skate *Bathyraja parmifera*, where they are exposed to fishing pressure. The main threat to skates from fisheries is incidental catch in longline and trawl vessels targeting groundfish such as Pacific Halibut and Pacific Cod (Ormseth, 2014a). In the USA, a short-lived target fishery for skates operated in the GOA from 2003 to 2005, eventually closing due to management concerns. The Big Skate dominates the skate biomass in waters that are <100 m deep in the GOA, while Longnose Skate dominates the skate biomass from 100 to 200 m depth (Ormseth, 2014a). In the EBS a single species, *B. parmifera*, dominates the shelf skate complex at depths of 50–200 m (Stevenson, 2004). The incidental catch of skates in the EBS and the northern part of the GOA is managed by the United States. Species-specific bottom trawl survey biomass estimates have been generated by the National Oceanic and Atmospheric Administration (NOAA) Fisheries every 3 years from 1984 to 1999, and surveys are now conducted every 2 years. In the eastern EBS, the Alaska Skate has separate harvest specifications from "other skates", which includes all other skate species caught (Ormseth, 2014b). In the GOA, the Big Skate and Longnose Skate have separate harvest specifications from "other skates", a general grouping that includes the Aleutian Skate *Bathyraja aleutica*, the Bering Skate *Bathyraja interrupta*, and the Alaska Skate, among others (Ormseth, 2014a). Species-specific management for the Big and Longnose skates is also in place in British Columbian waters—the only Canadian province on the Pacific Ocean. Off the coast of the contiguous United States skate fisheries are managed and assessed by the Pacific Fishery Management Council but skates are not a target species and do not represent a substantial portion of bottom trawl or longline catches (Gertseva and Schirripa, 2008).

The chondrichthyan fauna is slightly more diverse in the GOA compared to the EBS, as the southern region includes vagrants and species that are more common in the CAC. The northern region of the GOA, like the EBS, is skate-dominated and managed by NOAA Fisheries, while the southern region is more diverse and managed by the Department of Fisheries and Oceans (DFO) Canada. The North Pacific Spiny Dogfish *Squalus suckleyi* is among the most commonly encountered chondrichthyan species in the Canadian waters of the southern GOA (McFarlane et al., 2010). Despite having one of the lowest maximum intrinsic rates of population increase among all shark species, a target fishery for the North Pacific Spiny Dogfish has operated in British Columbian waters since 1870. Recent low demand for this species means that annual landings are well below the total allowable catch (Gallucci et al., 2011). This fishery has been touted as the quintessential example of a sustainable shark fishery and was granted Marine Stewardship Council certification in 2011, a status that was maintained following a 2012 audit. This species is currently assessed as Least Concern on the IUCN Red List for Threatened Species (Bigman et al., 2016), while its congener *Squalus acanthias* has been assessed as Vulnerable globally (Fordham et al., 2016) and Endangered in European waters (Ellis et al., 2015).

The large, filter-feeding Basking Shark *Cetorhinus maximus* historically ranged from the southern GOA to the northern CAC in NEP waters, aggregating in the waters off British Columbia in the summer and fall, and migrating to Californian waters in the winter and spring. This species has one of the lowest maximum intrinsic rates of population increase among all shark species (Pardo et al., 2016; Pauly, 2002), and target and bycatch fisheries, ship strikes, as well as a DFO eradication program between 1945 and 1970 have contributed to drastic reductions in the population. The eradication program, which included the use of blade-equipped boats for ramming sharks, was motivated by nuisance interactions between the Basking Shark and vessels gillnetting and trolling for salmon. The North Pacific subpopulation of this species is listed as Endangered on the IUCN Red List for Threatened Species (Fowler, 2009), is considered Endangered by the Committee on the Status of Endangered Wildlife in Canada (COSEWIC, 2007), is included under the Species At Risk Act in Canada, and is considered a Species of Concern by the Endangered Species Act in the United States. In order to gather information on this rare species, the Spot a Basking Shark Project was developed as a partnership between the Pacific Shark Research Center (Moss Landing Marine Laboratories) and NOAA Fisheries in 2010. Since the beginning of the project, 82 sightings have been reported; however, some of these sightings date back to 1990 (K. Walovich, Pacific Shark Research Center, Moss Landing Marine Laboratories, pers. comm.).

The second most threatened family of shark in the world are the angel sharks (Squatinidae), a coastal and continental shelf-dwelling family of flat-bodied sharks that have been driven locally or regionally extinct in many regions (Barker et al., 2016; Dulvy et al., 2003). Three species in the

Northeast Atlantic and Mediterranean are all considered Critically Endangered on the IUCN Red List for Threatened Species, and all four species in Taiwanese waters are threatened (Ebert et al., 2013a,b). A single species, the Pacific Angelshark *Squatina californica* occurs in the CAC and as a vagrant in the GOA. A commercial fishery in California beginning in the late 1970s historically targeted this species. During the expansion phase of this fishery landings grew rapidly, peaking in 1987 and beginning to decline in the early 1990s. It was suspected that this decline in landings may have been partially driven by overexploitation, yet management changes such as a 1989 minimum size limit and a 1994 ban on gill and trammel nets likely also contributed (Leet et al., 2001). Implementation of the MLPA initiative in the CAC was expected to have a positive effect on Pacific Angelshark populations, as well as several other species of chondrichthyan. This species is no longer targeted in Californian waters and is considered Least Concern in the USA on the IUCN Red List for Threatened Species. The Pacific Angelshark is still targeted in Mexican waters, where there is evidence that populations have declined and range has contracted, and this species is therefore considered Near Threatened globally (Cailliet et al., 2016).

14. CONSERVATION

The IUCN Species Survival Commission Shark Specialist Group evaluated the conservation status of all chondrichthyans and found that an estimated and predicted one-quarter are threatened with extinction (listed as Critically Endangered, Endangered, or Vulnerable on the IUCN Red List for Threatened Species; Dulvy et al., 2014a). The number of threatened species varies regionally—in the Mediterranean Sea 53% of chondrichthyans are considered threatened (Dulvy et al., 2016), while in the high-latitude seas only 14% of chondrichthyans are considered threatened (Ebert and Winton, 2010).

Within the NEP region, nearly 40% of the species that inhabit the area seasonally, or pass through during migration (transient species), have been assessed as threatened (Table 5). The majority of these transient species are wide ranging and threatened by fisheries where they are taken as targeted or incidental catch (i.e. the Oceanic Whitetip *Carcharhinus longimanus*, the Scalloped Hammerhead Shark *Sphyrna lewini*, and Smooth Hammerhead Shark *Sphyrna zygaena*). Of those that are threatened with extinction, two species are listed as Endangered (the Whale Shark *Rhincodon typus* and the Scalloped Hammerhead Shark) and 13 are listed as Vulnerable. Globally,

Table 5 Number of NEP Chondrichthyans Grouped by Their General Habitat and IUCN Red List Assessment Category

			IUCN Red List Assessment Category					
Habitat	**Species (%)**	**Threatened**	**CR**	**EN**	**VU**	**NT**	**LC**	**DD**
Coastal	553 (47.0)	130 (23.6)	16 (2.9)	28 (5.1)	86 (15.6)	81 (14.6)	138 (25.0)	204 (36.9)
Deep sea	542 (46.1)	27 (5.0)	1 (0.2)	9 (1.7)	17 (3.1)	43 (7.9)	171 (31.5)	301 (55.5)
Pelagic/Oceanic	50 (4.3)	17 (34.0)	0	3 (6.0)	14 (28.0)	13 (26.0)	14 (28.0)	6 (12.0)
Freshwater	31 (2.6)	13 (41.9)	3 (9.7)	7 (22.6)	3 (9.7)	1 (3.2)	1 (3.2)	16 (51.6)
Total	1176	187 (15.9)	20 (1.7)	47 (4.0)	120 (10.2)	138 (11.7)	324 (27.6)	527 (44.8)

The number of species considered to be Threatened is the sum total of Critically Endangered (CR), Endangered (EN), and Vulnerable (VU) species. The number of species in each of these three categories is also included, as well as the number of Near Threatened (NT), Least Concern (LC), and Data Deficient (DD) species. All IUCN Red List assessments, including archived assessments, are available online: www.iucnredlist.org.

nearly half of chondrichthyans are considered Data Deficient, meaning that information is insufficient to assess their status, while only about 20% of transient NEP species were Data Deficient. Of Data Deficient transient NEP species, the Broadnose Sevengill Shark *Notorynchus cepedianus*, Pacific Sharpnose Shark *Rhizoprionodon longurio*, and the Pacific Sleeper Shark *Somniosus pacificus* were predicted to be threatened based on body size and geographic distribution, while the other five Data Deficient transient species were predicted to be Least Concern (Dulvy et al., 2014a). Most of these transient species, except for the Common Thresher and Shortfin Mako Shark, are uncommon to rare in southern California waters. The Smooth Hammerhead Shark is somewhat common in southern California, while the Scalloped Hammerhead and Tiger Shark are rare and generally only seen in El Niño years.

In contrast, none of the chondrichthyans that have ranges mostly confined to the NEP (regional endemic species) are currently considered threatened with extinction (Table 6). These NEP regional endemic chondrichthyans were recently reassessed in 2015–2016, approximately a decade after their previous assessments (2003–2009). Over the decade, the number of Data Deficient species dropped, with all four of the species considered Data Deficient in 2003–2009 being reassessed as Least Concern

Table 6 Current and Previous IUCN Red List of Threatened Species Assessment Categories for Species That Have Most of Their Range Within NEP Waters (Regional Endemic Species) and Species That Are Transient in NEP Waters (Transient Species)

	Regional Endemic Species Assessments		**Transient Species Assessments**
IUCN Red List Assessment Category	**Current Status (2015–2016)**	**Previous Status (2003–2009)**	**Current Status**
Critically Endangered	0	0	0
Endangered	0	0	2
Vulnerable	0	0	13
Near Threatened	2	4	6
Least Concern	32	25	9
Data Deficient	5	9	8
Total	39	38	38

The dates that IUCN Red List assessments were conducted for regional endemic species are shown to allow for comparison. All IUCN Red List assessments, including archived assessments, are available online: www.iucnredlist.org.

in 2015–2016. The number of Near Threatened species was also reduced from four in 2003–2009 to two in 2014–2015. While changes in threat categories should be interpreted with caution—species can change categories as a result of improved knowledge that does not reflect a true change in threat status—the two changes here reflected genuine improvement in conservation status. In the case of the Big Skate *Beringraja binoculata*, the change in status from Near Threatened to Least Concern was supported by stock assessments that showed stable population sizes in some regions, and by the incorporation of this species into management plans in other regions (Farrugia et al., 2016). Of the five Data Deficient NEP regional endemic species, two species were predicted to be threatened, the Horn Shark *H. francisci* and the Banded Guitarfish *Zapteryx exasperata* (Dulvy et al., 2014a). Both of these species are restricted to the southern portion of the CAC, where they may be more exposed to targeted elasmobranch fisheries in Mexico.

15. SUMMARY AND CONCLUSIONS

The two leading threats to the world's chondrichthyans—overfishing and habitat loss—have been mitigated with much success in the NEP through fisheries management and habitat protection. While almost 40% of species that are transient in NEP waters have been assessed as threatened, threats to these species are mostly due to fisheries in waters outside of the NEP. None of the species considered to be regionally endemic to the NEP have been assessed as threatened, although some Data Deficient species that have ranges that extend into the southern half of the CAC may be threatened based on their body size and geographic distribution. Skates are the dominant chondrichthyan fauna of the EBS and GOA, and careful fisheries management means that all species are considered Least Concern or Data Deficient in these waters. Historical target fisheries for the Basking Shark severely depleted the population, which is still recovering and considered Endangered; while despite historical and current fisheries for the North Pacific Spiny Dogfish and historical fisheries for the Pacific Angelshark, both are considered Least Concern, despite having life histories that make them vulnerable to overfishing. Though new chondrichthyan species are being discovered rapidly by science, management infrastructure in the NEP seems well poised to accommodate these new discoveries and allow for their conservative management.

ACKNOWLEDGEMENTS

We would like to thank our colleagues Joe Bizzarro, Wade Smith, and Greg Cailliet, and all of the former and current students at the Pacific Shark Research Center/Moss Landing Marine Laboratories who have, and continue to, contribute to the advancement of Northeast Pacific Chondrichthyan research. We also thank Nick Dulvy, who supported the participation of Jennifer Bigman and Julia Lawson during the creation of this chapter. We also thank Chris Lowe, Connor White, Dayv Lowry, and Shawn Larson for their careful review of this chapter, which greatly improved the overall quality of the manuscript.

APPENDIX A. CHECKLIST OF NEP CHONDRICHTHYANS

Order Hexanchiformes

Family Chlamydoselachidae Garman, 1884. Frilled Sharks

Genus *Chlamydoselachus* Garman, 1884. Frilled Sharks

Chlamydoselachus anguineus Garman, 1884. Frilled Shark

Distribution: The frilled shark is known from a single specimen caught 22 miles off Point Arguello, California. The specimen was caught in a drift gill net at a depth of 20 m in water over 1500 m deep. Globally, this is a wide-ranging species, but with a scattered distribution.

IUCN Conservation status: Least Concern.

Family Hexanchidae Gray, 1851. Cowsharks

Genus *Hexanchus* Rafinesque, 1810. Sixgill Sharks

Hexanchus griseus (Bonnaterre, 1788). Bluntnose Sixgill Shark

Distribution: A common deep-water shark found throughout the NEP at depth, from Alaskan waters to southern California. Wide ranging with a circumglobal distribution.

IUCN Conservation status: Near Threatened.

Genus *Notorynchus* Ayres, 1855. Broadnose Sevengill Sharks

Notorynchus cepedianus (Péron, 1807). Broadnose Sevengill Shark

Distribution: Southeast Alaska to southern California, very common in several northern California bays and seasonally abundant in Willapa Bay and Grays Harbor, WA. Globally, this species occurs in most temperate seas.

IUCN Conservation status: Data Deficient.

Order Echinorhiniformes

Family Echinorhinidae Gill, 1862. Bramble Sharks

Genus *Echinorhinus* Blainville, 1816. Bramble Sharks

Echinorhinus cookei Pietschmann, 1928. Prickly Shark

Distribution: Moolach Beach, Oregon to southern California, but occurs in the eastern Pacific to at least Peru. Elsewhere, sporadically distributed throughout the Pacific Ocean, and possibly in the eastern Indian Ocean.

IUCN Conservation status: Near Threatened.

Order Squaliformes

Family Squalidae Blainville, 1816. Dogfishes

Genus *Squalus* Linnaeus, 1758. Dogfishes

Squalus suckleyi (Girard, 1854). North Pacific Spiny Dogfish

Distribution: Found throughout the North Pacific, with some individually occasionally found in the southern Bering Sea, but very common from the Gulf of Alaska to southern California. Known to migrate from the NEP to the Northwest Pacific.

IUCN Conservation status: Least Concern.

Family Etmopteridae Fowler, 1934. Lanternsharks/Tiburones Luceros

Genus *Centroscyllium* Müller and Henle, 1841. Combtooth Dogfishes

Centroscyllium nigrum Garman, 1899. Pacific Black Dogfish

Distribution: Southern California and southwards to Chile in the eastern Pacific Ocean.

IUCN Conservation status: Data Deficient.

Family Somniosidae Jordan, 1888. Sleeper Sharks

Genus *Somniosus* Lesueur, 1818. Sleeper Sharks

Somniosus pacificus Bigelow and Schroeder, 1944. Pacific Sleeper Shark

Distribution: Bering Sea and Gulf of Alaska, and to southern California, this is a wide-ranging deep-sea species found throughout the North Pacific Ocean. Its occurrence towards to equator and into the southern hemisphere is uncertain due to misidentification with the Southern Sleeper Shark *Somniosus antarcticus*.

IUCN Conservation status: Data Deficient.

Genus *Zameus* Jordan and Fowler, 1903. Velvet Dogfishes

Zameus squamulosus (Günther, 1877). Velvet Dogfish

Distribution: Recently discovered to occur off southern California, this wide-ranging species is most common in warm-temperate and tropical seas.

IUCN Conservation status: Data Deficient.

Family Dalatiidae Gray, 1851. Kitefin Sharks

Genus *Euprotomicrus* Gill, 1865. Pygmy Sharks

Euprotomicrus bispinatus (Quoy and Gaimard, 1824). Pygmy Shark

Distribution: This oceanic species is known from a few individuals taken nearly 500 miles off southern California. A cosmopolitan species occurring mostly in the open ocean of all three major oceans.

IUCN Conservation status: Least Concern.

Genus *Isistius* Gill, 1865. Cookiecutter Sharks

Isistius brasiliensis (Quoy and Gaimard, 1824) Cookiecutter Sharks

Distribution: The presence of this species in the NEP was only recently confirmed based on a couple of individuals caught off southern California. However, crater wounds on pinnipeds have suggested its occurrence for some time. A wide-ranging species in most warm-temperate to tropical seas.

IUCN Conservation status: Least Concern.

Order Squatiniformes

Family Squatinidae Bonaparte, 1838. Angel Sharks

Genus *Squatina* Duméril, 1806. Angel Sharks

Squatina californica Ayres, 1859. Pacific Angelshark

Distribution: Endemic to the eastern Pacific from southeastern Alaska to the Gulf of California. Records of its occurrence further to the south require confirmation.

IUCN Conservation status: Near Threatened.

Order Heterodontiformes

Family Heterodontidae Gray, 1851. Bullhead Sharks

Genus *Heterodontus* Blainville, 1816. Bullhead Sharks

Heterodontus francisci (Girard, 1855). Horn Shark

Distribution: Endemic to the eastern Pacific from southern California to the Gulf of California.

IUCN Conservation status: Data Deficient.

Order Orectolobiformes

Family Rhincodontidae Müller and Henle, 1839. Whale Shark

Genus *Rhincodon* Smith, 1829. Whale Shark

Rhincodon typus (Smith, 1828). Whale Shark

Distribution: Small numbers of these sharks are periodically observed off the California coast, mostly from central California and southwards, but there is one record from off northern California. A circumglobal species, the Whale Shark is found in warm-temperate and tropical seas.

IUCN Conservation status: Endangered.

Order Lamniformes

Family Mitsukurinidae Jordan, 1898. Goblin Shark

Genus *Mitsukurina* Jordan, 1898. Goblin Shark

Mitsukurina owstoni Jordan, 1898. Goblin Shark

Distribution: Known from a single specimen caught off San Clemente Island, off southern California. Elsewhere, a wide-ranging deep-sea species with a scattered, but global distribution.

IUCN Conservation status: Least Concern.

Family Odontaspididae Müller and Henle, 1839. Sandtiger Sharks

Genus *Odontaspis* Agassiz, 1838. Deep-sea Sandtiger Sharks

Odontaspis ferox (Risso, 1810). Smalltooth Sandtiger Sharks

Distribution: This species is known from about a half-dozen records, all from off southern California. It is wide ranging and has been reported in all three major oceans.

IUCN Conservation status: Vulnerable.

Family Megachasmidae Taylor, Compagno, and Struhsaker, 1983. Megamouth Shark

Genus Megachasma Taylor, Compagno, and Struhsaker, 1983. Megamouth Shark

Megachasma pelagios Taylor, Compagno, and Struhsaker, 1983. Megamouth Shark

Distribution: First discovered off Hawaii in 1976, most records have come from the western North Pacific, but southern California and northern Baja California is an area where more than a half-dozen individuals have been observed.

IUCN Conservation status: Least Concern.

Family Alopiidae Bonaparte, 1838. Thresher Sharks

Genus *Alopias* Rafinesque, 1810. Thresher Sharks

Alopias pelagicus Nakamura, 1935. Thintail or Pelagic Thresher Shark

Distribution: Observed off southern California during warm water years, this species is found throughout the warm-temperate and tropical Pacific and Indian oceans.

IUCN Conservation status: Vulnerable.

Alopias superciliosus Lowe, 1841. Bigeye Thresher Shark

Distribution: Mostly observed off southern California, this species occasionally ranges as far north as British Columbia. Globally, this species occurs in most warm-temperate to tropical seas.

IUCN Conservation status: Vulnerable.

Alopias vulpinus (Bonnaterre, 1788). Common Thresher Shark

Distribution: British Columbia to southern California, there is one unconfirmed record of this species from southeast Alaska; it is most common from California and southwards. Wide ranging in most temperate and tropical seas.

IUCN Conservation status: Vulnerable.

Family Cetorhinidae Gill, 1862. Basking Shark

Genus *Cetorhinus* Blainville, 1816. Basking Shark

Cetorhinus maximus (Gunnerus, 1765). Basking Shark

Distribution: Southern Bering Sea, Alaska to southern California, this is a wide-ranging species, found in all major oceans, known to make long-range migrations through tropical seas, but at depth where the waters are cooler.

IUCN Conservation status: Vulnerable.

Family Lamnidae Müller and Henle, 1838. Mackerel Sharks

Genus *Carcharodon* Smith *in* Müller and Henle, 1838. Great White Shark

Carcharodon carcharias (Linnaeus, 1758). Great White Shark

Distribution: Found from the Gulf of Alaska to southern California, they appear to be most abundant off California seasonally. A cosmopolitan species found in most temperate to tropical seas.

IUCN Conservation status: Vulnerable.

Genus *Isurus* Rafinesque, 1810. Mako Sharks

Isurus oxyrinchus Rafinesque, 1810. Shortfin Mako Shark

Distribution: Most common from Oregon to southern California, but this wide-ranging species is common in most temperate and tropical seas.

IUCN Conservation status: Vulnerable.

Isurus paucus Guitart, 1966. Longfin Mako Shark

Distribution: Known from a few records off southern California, it may be more common, but has been misidentified with the more common Shortfin Mako Shark. The Longfin Mako Shark is sporadically distributed in all major oceans.

IUCN Conservation status: Vulnerable.

Genus *Lamna* Cuvier, 1816. Porbeagle Sharks

Lamna ditropis Hubbs and Follet, 1947. Salmon Shark

Distribution: Wide ranging throughout the North Pacific, it is most common in the NEP from central California northwards to the Gulf of Alaska.

IUCN Conservation status: Least Concern.

Order Carcharhiniformes

Family Scyliorhinidae Gill, 1862

Genus *Apristurus* Garman, 1913

Apristurus brunneus (Gilbert, 1892). Brown Catshark

Distribution: Endemic to the eastern Pacific, this species occurs from southern British Columbia to Mexico and possibly to Peru.

IUCN Conservation status: Data Deficient.

Apristurus kampae (Taylor, 1972). Longnose Catshark

Distribution: Endemic to the eastern Pacific, this species occurs from Oregon to Peru.

IUCN Conservation status: Data Deficient.

Genus *Cephaloscyllium* Gill, 1862

Cephaloscyllium ventriosum (Garman, 1880). Swellshark

Distribution: Endemic to the eastern Pacific, this species occurs from central California to Chile.

IUCN Conservation status: Least Concern.

Genus *Parmaturus* Garman, 1906

Parmaturus xaniurus (Gilbert, 1892). Filetail Catshark

Distribution: Endemic to the eastern Pacific, this species occurs from Oregon to the Gulf of California.

IUCN Conservation status: Least Concern.

Family Triakidae Gray, 1851

Genus *Galeorhinus* Blainville, 1816

Galeorhinus galeus (Linnaeus, 1758). Soupfin Shark

Distribution: British Columbia to Baja California, it is found temperate seas of the South Pacific, South Atlantic, southwestern Indian Ocean, and Northeastern Atlantic.

IUCN Conservation status: Vulnerable.

Genus *Mustelus* Linck, 1790

Mustelus californicus Gill, 1864. Grey Smooth-hound Shark

Distribution: Endemic to the eastern Pacific, this species occurs from California to Mazatlán, Mexico, including the Gulf of California.

IUCN Conservation status: Least Concern.

Mustelus henlei (Gill, 1863). Brown Smooth-hound Shark

Distribution: Endemic to the eastern Pacific, this species occurs from Washington to the Gulf of California, and from Ecuador and Peru.

IUCN Conservation status: Least Concern.

Mustelus lunulatus Jordan and Gilbert, 1882. Sicklefin Smooth-hound Shark

Distribution: Endemic to the eastern Pacific, this species occurs from California to the Gulf of Panama.

IUCN Conservation status: Least Concern.

Genus *Triakis* Müller and Henle, 1838

Triakis semifasciata (Girard, 1855). Leopard Shark

Distribution: Distribution: Endemic to the eastern Pacific, this species occurs from Washington to Mazatlán, Mexico, including the Gulf of California.

IUCN Conservation status: Least Concern.

Family Carcharhinidae Jordon and Evermann, 1896

Genus *Carcharhinus* Blainville, 1816

Carcharhinus brachyurus (Günther, 1870). Copper Shark

Distribution: A seasonal visitor to southern California from Baja California during the warm water summer months. Elsewhere, this cosmopolitan species occurs in warm-temperate to subtropical seas.

IUCN Conservation status: Near Threatened.

Carcharhinus longimanus (Poey, 1861). Oceanic Whitetip Shark

Distribution: A rare species in southern California waters, usually observed around the Channel Islands during warm water years. A cosmopolitan species, this is one of the most common oceanic sharks in tropical and warm-temperate seas.

IUCN Conservation status: Vulnerable.

Carcharhinus obscurus (Lesueur, 1818). Dusky Shark

Distribution: Southern California, particularly during warm water years, this wide-ranging species is common in most warm-temperate and tropical seas.

IUCN Conservation status: Vulnerable.

Genus *Galeocerdo* Müller and Henle, 1838

Galeocerdo cuvier (Péron and Lesueur *in* Lesueur, 1822). Tiger Shark

Distribution: Observed occasionally during warm water years, the Tiger Shark is most abundant in tropical seas around the world.

IUCN Conservation status: Near Threatened.

Genus *Prionace* Cantor, 1849

Prionace glauca (Linnaeus, 1758). Blue Shark

Distribution: The Blue Shark is one of the most wide-ranging species globally and is found in the NEP from the Gulf of Alaska, where it is uncommon, to southern California where it is very abundant seasonally.

IUCN Conservation status: Near Threatened.

Genus *Rhizoprionodon* Whitley, 1929

Rhizoprionodon longurio (Jordan and Gilbert, 1882). Pacific Sharpnose Shark

Distribution: A warm-temperate species occasionally observed in southern California waters, it is common further south and into the Gulf of California.

IUCN Conservation status: Data Deficient.

Family Sphyrnidae

Genus *Sphyrna* Rafinesque, 1810

Sphyrna lewini (Griffith and Smith *in* Cuvier, Griffith and Smith, 1834). Scalloped Hammerhead Shark

Distribution: Hammerhead sharks can be difficult to distinguish, but the Scalloped Hammerhead Shark is most commonly observed in tropical and warm-temperate waters. It occasionally wonders into southern California during warm waters years, but is a rare visitor.

IUCN Conservation status: Endangered.

Sphyrna tiburo (Linnaeus, 1758). Bonnethead Shark

Distribution: A common eastern Pacific warm-temperate to tropical species that has been recorded from off San Diego in southern California, but this species is rare in the NEP.

IUCN Conservation status: Least Concern.

Sphyrna zygaena (Linnaeus, 1758). Smooth Hammerhead Shark

Distribution: This is the most cool-temperate species of hammerhead sharks and is the most common hammerhead shark species observed in southern California waters.

IUCN Conservation status: Vulnerable.

Order Torpediniformes

Family Platyrhinidae Müller and Henle, 1838

Genus *Platyrhinoidis* Garman, 1881

Platyrhinoidis triseriata (Jordan and Gilbert, 1880). Thornback Ray

Distribution: Endemic to the eastern Pacific, this species occurs from central California to the Gulf of California.

IUCN Conservation status: Least Concern.

Family Torpedinidae Bonaparte, 1838

Genus *Tetronarce* Gill, 1862

Tetronarce californica (Ayres, 1855). California Electric Ray

Distribution: Endemic to the eastern Pacific, this species occurs from British Columbia to Mazatlán, Mexico.

IUCN Conservation status: Least Concern.

Order Rhinopristiformes

Family Rhinobatidae Müller and Henle, 1837

Genus *Pseudobatos*

Pseudobatos productus (Ayres, 1854). Shovelnose Guitarfish

Distribution: Endemic to the eastern Pacific, this species occurs from central California to the Gulf of California, and possibly Mazatlán, Mexico.

IUCN Conservation status: Near Threatened.

Family Trygonorrhinidae Last, Séret, and Naylor, 2016

Genus *Zapteryx* Jordan and Gilbert, 1880

Zapteryx exasperata (Jordan and Gilbert, 1880). Banded Guitarfish

Distribution: Endemic to the eastern Pacific, this species occurs from southern California to about Mazatlán, Mexico, where it is replaced by a different species, the Spotted Guitarfish *Zapteryx xyster*.

IUCN Conservation status: Data Deficient.

Order Rajiformes

Family Arhynchobatidae Fowler, 1934

Genus *Bathyraja* Ishyama, 1958

Bathyraja abyssicola (Gilbert, 1896). Deepsea Skate

Distribution: Eastern Pacific from the northern Bering Sea to northern Baja California off Coronado Island and Cortes Bank. It has also been reported as far west as Japan.

IUCN Conservation status: Data Deficient.

Bathyraja aleutica (Gilbert, 1896). Aleutian Skate

Distribution: Endemic to the North Pacific, ranging from northern Japan and the Bering Sea to off Cape Mendocino, California.

IUCN Conservation status: Least Concern.

Bathyraja interrupta (Gill and Townsend, 1897). Bering Skate

Distribution: Endemic to the eastern Pacific, from the Bering Sea to the Gulf of Alaska, and British Columbia. Its exact distribution is uncertain due to confusion with the Sandpaper Skate *Bathyraja kincaidii*; these two species may in fact represent a species complex involving several species.

IUCN Conservation status: Least Concern.

Bathyraja kincaidii (Garman, 1908). Sandpaper Skate

Distribution: Endemic to the NEP, ranging from British Columbia to northern Baja California.

IUCN Conservation status: Data Deficient.

Bathyraja lindbergi Ishiyama and Ishihara, 1977. Commander Skate

Distribution: Endemic to the North Pacific from Japan and the Bering Sea to British Columbia.

IUCN Conservation status: Least Concern.

Bathyraja maculata Ishiyama and Ishihara, 1977. White-blotched Skate

Distribution: Endemic to the North Pacific from Japan and the Bering Sea, including the Aleutian Islands, and into the Gulf of Alaska, where it is rare, to Southeast Alaska.

IUCN Conservation status: Least Concern.

Bathyraja mariposa Stevenson et al., 2004. Butterfly Skate

Distribution: Endemic to the Central Aleutian Islands.

IUCN Conservation status: Least Concern.

Bathyraja microtrachys (Osburn and Nichols, 1916). Fine-spine Skate

Distribution: One of the deepest occurring skates, to at least 3300 m, this species is endemic to the NEP, ranging from off San Diego, California to Washington.

IUCN Conservation status: Least Concern.

Bathyraja minispinosa Ishiyama and Ishihara, 1977. Smallthorn Skate

Distribution: Endemic to the Bering Sea, Aleutian Islands, and westwards to Japan.

IUCN Conservation status: Least Concern.

Bathyraja panthera Stevenson, Orr, Hoff, Spies, and McEachran, 2011. Leopard Skate

Distribution: Endemic to the western Aleutian Islands.

IUCN Conservation status: Least Concern.

Bathyraja parmifera (Bean, 1881). Alaska Skate

Distribution: Endemic to the Gulf of Alaska, Bering Sea, Aleutian Islands, and northern Japan.

IUCN Conservation status: Least Concern.

Bathyraja spinosissima (Beebe and Tee-Van, 1941). Pacific White Skate

Distribution: Eastern Pacific from Waldport, Oregon to the Galapagos Islands, also reported from the Sea of Okhotsk off northern Japan.

IUCN Conservation status: Least Concern.

Bathyraja taranetzi Dolganov, 1983. Mud Skate

Distribution: Bering Sea and Aleutian Islands, and westwards to the northern Kuril Islands.

IUCN Conservation status: Least Concern.

Bathyraja trachura (Gilbert, 1892). Roughtail Skate

Distribution: Northern Baja California to the Bering Sea, and to the Sea of Okhotsk.

IUCN Conservation status: Least Concern.

Bathyraja violacea (Survorov, 1935). Okhotsk Skate

Distribution: Aleutian Islands, and western Bering Sea to Japan.

IUCN Conservation status: Data Deficient.

Family Rajidae Blainville, 1816

Genus *Amblyraja* Malm, 1877

Amblyraja badia (Garman, 1899). Broad Skate

Distribution: Northern Bering Sea to Panama, and possibly from off northern Japan.

IUCN Conservation status: Least Concern.

Genus *Beringraja* Ishihara, Treloar, Bor, Sensou, and Jeong, 2012

Beringraja binoculata (Girard, 1855). Big Skate

Distribution: Southeastern Bering Sea where it is uncommon to Cedros Island, central Baja California.

IUCN Conservation status: Least Concern.

Genus *Raja* Linnaeus, 1758

Raja inornata Jordan and Gilbert, 1881. California Skate
Distribution: Southern British Columbia to central Baja California, with a disjunct population in the northern Gulf of California.

IUCN Conservation status: Least Concern.

Raja rhina Jordan and Gilbert, 1880. Longnose Skate
Distribution: Southeastern Bering Sea where it is uncommon to Cedros Island, central Baja California, and the Gulf of California.

IUCN Conservation status: Least Concern.

Raja stellulata Jordan and Gilbert, 1880. Starry Skate
Distribution: Southern British Columbia to northern Baja California.

IUCN Conservation status: Least Concern.

Order Myliobatiformes

Family Urotrygonidae

Genus *Urobatis* Garman, 1913

Urobatis halleri (Cooper, 1863). Round Stingray
Distribution: Endemic to the eastern Pacific from northern California to Panama.

IUCN Conservation status: Least Concern.

Family Dasyatidae Jordan, 1888

Genus *Hypanus*

Hypanus dipterurus (Jordan and Gilbert, 1880). Diamond Stingray
Distribution: Southern California to northern Chile, and possibly the Galapagos Islands.

IUCN Conservation status: Data Deficient.

Genus *Pteroplatytrygon* Fowler, 1910

Pteroplatytrygon violacea (Bonaparte, 1832). Pelagic Stingray
Distribution: Eastern Pacific from British Columbia to central Chile. A wide-ranging species found in most warm-temperate and tropical seas.

IUCN Conservation status: Least Concern.

Family Gymnuridae Fowler, 1934

Genus *Gymnura* van Hasselt, 1823

Gymnura marmorata (Cooper, 1864). California Butterfly Ray
Distribution: Endemic to the eastern Pacific from southern California to Peru.

IUCN Conservation status: Least Concern.

Family Myliobatidae Bonaparte, 1838

Genus *Myliobatis* Cuvier, 1816

Myliobatis californica Gill, 1865. Bat Ray

Distribution: Endemic to the eastern Pacific from northern Oregon to the Gulf of California.

IUCN Conservation status: Least Concern.

Family Mobulidae Gill, 1829

Genus *Manta* Bancroft, 1829

Manta birostris (Walbaum, 1792). Pelagic Manta

Distribution: Southern California, where they are uncommon, but this wide-ranging species is common in warm-temperate and tropical waters.

IUCN Conservation status: Vulnerable.

Genus *Mobula* Rafinesque, 1810

Mobula japonica (Müller and Henle, 1841). Spinetail Mobula

Distribution: Southern California to Peru, but common in most warm-temperate and tropical seas.

IUCN Conservation status: Near Threatened.

Order Chimaeriformes

Family Chimaeridae Rafinesuqe, 1815

Genus *Hydrolagus* Gill, 1862

Hydrolagus colliei (Lay and Bennett, 1839). White-spotted Ratfish

Distribution: Endemic to the eastern Pacific from Alaska to Central America.

IUCN Conservation status: Least Concern.

Hydrolagus melanophasma James et al., 2009. Eastern Pacific Black Ghost Shark

Distribution: Endemic to the eastern Pacific from central California to Chile.

IUCN Conservation status: Least concern.

Hydrolagus cf. *trolli*

Distribution: Wide ranging in the southern hemisphere, but known from remote-operated vehicle video observations from off central and southern California, and Hawaii.

IUCN Conservation status: Least Concern.

Family Rhinochimaeridae Garman, 1901

Genus *Harriotta* Gill, 1862

Harriotta raleighana (Goode and Bean, 1895). Longnose Chimaera

Distribution: Eastern Pacific from southern California to Peru, but also with a scattered, circumglobal distribution.

IUCN Conservation status: Least Concern.

APPENDIX B. KEY TO NORTHEASTERN PACIFIC CHONDRICHTHYANS (MODIFIED FROM EBERT, 2003)

1a. One paired gill opening. 2. (Chimaeriformes)

1b. Five to seven paired gill openings. 5.

2a. Snout elongated and tapering. Longnose Chimaera, *Harriotta raleighana* (Rhinochimaeridae)

2b. Snout short and blunt. 3. (Chimaeridae)

3a. Body surface with numerous, scattered white spots. Second dorsal fin with a deep notched, giving it the appearance of two fins. Spotted Ratfish, *Hydrolagus colliei*

3b. Body surface without spots or other markings. Second dorsal fin continuous, without a notch. 4.

4a. Dark brownish to black with light band across snout prior to preservation; uniformly dark coloured after preservation. Eastern Pacific Black Ghostshark, *Hydrolagus melanophasma*

4b. Light coloured, milky white with purplish tinge on fins in life. Pointy-nosed Blue Chimaera, *Hydrolagus* cf. *trolli*

5a. Body not ray-like. 6.

5b. Body flat, ray-like. 46.

6a. One dorsal fin, six or seven paired gill openings. 7. (Hexanchiformes)

6b. Two dorsal fins, five paired gill openings. 9.

7a. Body eel-like, six paired gill slits with the first extending across the throat. Frilled Shark, *Chlamydoselachus anguineus* (Chlamydoselachidae)

7b. Body fusiform, six or seven paired gill slits without the first extending across the throat. 8. (Hexanchidae)

8a. Six paired gill slits. Bluntnose Sixgill Shark, *Hexanchus griseus*

8b. Seven paired gill slits. Broadnose Sevengill Shark, *Notorynchus cepedianus*

9a. Anal fin absent. 10.

9b. Anal fin present. 16.

10a. First dorsal fin originating behind pelvic fins. Prickly Shark, *Echinorhinus cookei* (Echinorhiniformes: Echinorhinidae)

10b. First dorsal fin originating in front of pelvic fins. 11. (Squaliformes)

11a. Dorsal fins without fin spines; first dorsal fin originating closer to pelvic fins. 12. (Dalatiidae)

11b. Dorsal fins with fin spines; first dorsal fin originating closer to pectoral fins. 13.

12a. First dorsal fin inserting well anterior to pelvic fin bases; second dorsal fin base about four times the length of first. Pygmy Shark, *Euprotomicrus bispinatus*

12b. First dorsal fin inserting over pelvic fin bases; both dorsal fin bases are similar in size. Cookiecutter Shark, *Isistius brasiliensis*

13a. Teeth similar in both jaws. North Pacific Spiny Dogfish, *Squalus suckleyi* (Squalidae)

13b. Teeth dissimilar on upper and lower jaws. 14.

14a. Fin spine preceding each dorsal fin prominent, height about equal to or greater than fin height. Pacific Black Dogfish, *Centroscyllium nigrum* (Etmopteridae)

14b. Fin spine may be present or absent, if present relatively inconspicuous. 15. (Somniosidae)

15a. Dorsal fin spines absent. Pacific Sleeper Shark, *Somniosus pacificus*

15b. Dorsal fin spines present, though short and partly covered by skin. Velvet Dogfish, *Zameus squamulosus*

16a. Dorsal fin spines present. Horn Shark, *Heterodontus francisci* (Heterodontiformes)

16b. Dorsal fin spines absent. 17.

17a. Mouth located in front of eyes. Whale Shark, *Rhincodon typus* (Orectolobiformes: Rhincodontidae)

17b. Mouth located behind eye. 18.

18a. No nictitating eyelids. 19. (Lamniformes)

18b. Nictitating eyelids. 29. (Carcharhiniformes)

19a. Snout extremely elongated, flat, and blade-like. Goblin Shark, *Mitsukurina owstoni* (Mitsukurinidae)

19b. Snout short to moderately, not greatly, elongated or blade-like, and broadly rounded. 20.

20a. Caudal fin about as long as trunk of body. 21. (Alopiidae)

20b. Caudal fin much shorter than trunk of body. 23.

21a. Head with a deep horizontal groove extending around each side. Eyes very large, with orbits expanded onto the dorsal head surface. Bigeye Thresher, *Alopias superciliosus*

21b. Head without a deep horizontal groove extending around each side. Eyes small, with orbits not expanded onto the dorsal head surface. 22.

22a. Flanks above pectoral and pelvic fins dark. Head narrow and snout elongated. Labial furrows absent. Pelagic Thresher, *Alopias pelagicus*

22b. Flanks above pectoral and pelvic fins white. Head broad and snout short. Labial furrows present. Common Thresher Shark, *Alopias vulpinus*

23a. Mouth terminal, lower jaw extending to snout tip. Megamouth Shark, *Megachasma pelagios* (Megachasmidae)

23b. Mouth subterminal, lower jaw not extending to snout tip. 24.

24a. Upper precaudal pit present; lower pit and lateral keels absent; caudal fin heterocercal, not crescent-shaped. Smalltooth Sandtiger Shark, *Odontaspis ferox* (Odontaspididae)

24b. Upper and lower precaudal pits present; strong lateral keels absent; caudal fin crescent-shaped. 25.

25a. Teeth minute, hook-shaped, but not blade-like. Gill slits extremely large, dorsally extending onto surface of head and the first nearly extending across the throat, and with internal gill rakers. Basking Shark, *Cetorhinus maximus* (Cetorhinidae)

25b. Teeth large, triangular, and blade-like. Gill slits large, but not extending onto dorsal surface of head or nearly across throat, and without internal gill rakers. 26. (Lamnidae)

26a. Teeth broadly triangular and serrated. White Shark, *Carcharodon carcharias*

26b. Teeth long, narrow, and without serrations. 27.

27a. Teeth with small lateral cusplets, origin of first dorsal fin over base or inner margin of pectoral fin, and a secondary keel on caudal peduncle. Salmon Shark, *Lamna ditropis*

27b. Teeth without small lateral cusplets, origin of first dorsal fin posterior to rear tip of inner margin of pectoral fin, and lacking a secondary keel on caudal peduncle. 28.

28a. Underside of snout and mouth area white; cusps of anterior teeth recurved forward; pectoral fins much shorter than head length; anal fin originating below the second dorsal fin base. Shortfin Mako, *Isurus oxyrinchus*

28b. Underside of snout and mouth area dusky; cusps of anterior teeth straight; pectoral fins about equal to head length; anal fin originating below or posterior to the second dorsal fin insertion. Longfin Mako, *Isurus paucus*

29a. First dorsal fin base opposite or behind pelvic bases. 30. (Scyliorhinidae)

29b. First dorsal base in front of pelvic bases. 33.

30a. Supraorbital crests present on cranium, above eyes. Ability to swell body trunk into nearly spherical shape by swallowing water or air. Swellshark, *Cephaloscyllium ventriosum*

30b. Supraorbital crests absent from cranium. Lacks ability to swell body trunk into spherical shape by swallowing water or air. 31.

31a. Anterior margin of upper caudal fin with a prominent row of enlarged dermal denticles. Filetail Catshark, *Parmaturus xaniurus*

31b. Anterior margin of upper caudal fin without a prominent row of enlarged dermal denticles. 32.

32a. Upper labial furrows longer than lower. Posterior fin margins not white-edged. Anal fin long and low, nearly reaching base of the caudal fin. Brown Catshark, *Apristurus brunneus*

32b. Upper labial furrows not longer than lower. Posterior fin margins white-edged. Anal fin short and high, not reaching base of the caudal fin. Longnose Catshark, *Apristurus kampae*

33a. Head with lateral, blade-like expansions. 34. (Sphyrnidae)

33b. Head normal, without lateral expansions. 36.

34a. Head expanded into a distinctive, short-lobed bonnet shape. Bonnethead, *Sphyrna tiburo*

34b. Head expanded into a distinctive, long-lobed hammer shape. 35.

35a. Anterior margin of head with a prominent central indentation. Scalloped Hammerhead, *Sphyrna lewini*

35b. Anterior margin of head without a prominent central indentation. Smooth Hammerhead, *Sphyrna zygaena*

36a. Precaudal pits absent. Spiracles present. 37. (Triakidae)

36b. Precaudal pits present. Spiracles absent. 41. (Carcharhinidae)

37a. Well-defined, bold, black saddlebars and dark spots on the dorsal surface. Leopard Shark, *Triakis semifasciata*

37b. No saddlebars, dark spots, or other prominent markings on the dorsal surface. 38.

38a. Second dorsal fin is much smaller than first dorsal fin and about as large as anal fin, terminal lobe of caudal fin about half the length of the dorsal caudal fin margin. Soupfin Shark, *Galeorhinus galeus*

38b. Second dorsal fin is nearly as large as first dorsal fin and much larger than anal fin, terminal lobe of caudal fin less than half the length of the dorsal caudal fin margin. 39.

39a. Posterior margin of dorsal fins frayed. Brown Smooth-hound, *Mustelus henlei*

39b. Posterior margin of dorsal fins not frayed. 40.

40a. First dorsal fin originating behind pectoral fin, with posterior margin slightly inclined backward, lower lobe of caudal fin indistinct. Grey Smooth-hound, *Mustelus californicus*

40b. First dorsal fin originating over free rear tips of pectoral fins, with posterior margin vertical from fin apex, lower lobe elongated and pointed in larger specimens. Sicklefin Smooth-hound, *Mustelus lunulatus*

41a. Spiracles present; upper labial furrows elongated, extending to front of eyes; prominent lateral ridges on caudal peduncle. Tiger Shark, *Galeocerdo cuvier*

41b. Spiracles absent; upper labial furrows not elongated, or extending to front of eyes; no prominent lateral ridges on caudal peduncle. 42.

42a. Midpoint of first dorsal fin base closer to pelvic fin origin than to pectoral insertion; colour bright blue in life. Blue Shark, *Prionace glauca*

42b. Midpoint of first dorsal fin base is closer to pectoral fin insertion than to pelvic fin origin; colour is grey to bronze in life. 43.

43a. Anal fin origin well in front of second dorsal fin origin, but never past midpoint of anal fin base. Pacific Sharpnose Shark, *Rhizoprionodon longurio*

43b. Anal fin origin under or behind second dorsal fin origin, and always anterior to midpoint of anal fin base. 44.

44a. Interdorsal ridge absent. Copper Shark, *Carcharhinus brachyurus.*

44b. Interdorsal ridge present. 45.

45a. Dorsal and pectoral fins broadly rounded at apices, with fin tips mottled white or with some black markings, mostly prominent in juveniles. Oceanic Whitetip Shark, *Carcharhinus longimanus*

45b. Dorsal and pectoral fins not broadly rounded at apices, posterior margin of pectoral fins distinctly concave and fin tips not mottled white or black; colour is mostly grey with no distinctive markings. Dusky Shark, *Carcharhinus obscurus*

46a. Mouth terminal, gills lateral on head, pectoral fins not attached to head. Pacific Angelshark, *Squatina californica* (Squatiniformes: Squatinidae)

46b. Mouth and gills ventral, pectoral fins attached to head. 47.

47a. Caudal fin prominent. 48.

47b. Caudal fin small or absent. 51.

48a. Disc with kidney-shaped electric organs present. Pacific Torpedo Ray, *Tetronarce californica* (Torpediniformes: Torpedinidae)

48b. Disc without kidney-shaped electric organs present. 49.

49a. Head and pectoral fins forming heart-shaped disc; one to three rows of large hook-like thorns along midback; first dorsal fin closer to caudal fin than to pelvic fins. Thornback Ray, *Platyrhinoidis triseriata* (Torpediniformes: Platyrhinidae)

49b. Head and pectoral fins forming wedge-shaped disc; small, single row of thorns along midback; first dorsal fin closer to pelvic fins than to caudal fin. 50. (Rhinopristiformes)

50a. Disc length greater than width; dorsal surface usually plain, occasionally with dark spots, but without prominent dark bars on back. Shovelnose Guitarfish, *Rhinobatos productus* (Rhinobatidae)

50b. Disc length about equal to width; dorsal surface with several prominent dark bars on back. Banded Guitarfish, *Zapteryx exasperata* (Trygonorrhinidae)

51a. Tail base with no stinging spine, with two dorsal fins. 52. (Rajiformes)

51b. Tail base with stinging spine, one or no dorsal fins. 71. (Myliobatiformes)

52a. Snout soft and flexible. 53. (Arhynchobatidae)

52b. Snout rigid and stiff. 67. (Rajidae)

53a. Scapular thorns present. 54.

53b. Scapular thorns absent. 58.

54a. Orbital thorns present, but may be obscure; tail thorns surrounding dermal denticles absent. 55.

54b. Orbital thorns absent; tail thorns surrounding dermal denticles present. 56.

55a. Dorsal colour greenish brown with a pattern of black spots and yellow blotches; tail thorns 19–28. Leopard Skate, *Bathyraja panthera*

55b. Dorsal colour nearly uniform brown; tail thorns range 15–23. Alaska Skate, *Bathyraja parmifera*

56a. Dorsal surface with a discontinuous row of median thorns. Bering Skate *Bathyraja interrupta*

56b. Dorsal surface with a continuous row of median thorns. 57.

57a. Ventral surface is white except for snout, posterior disc margins, pelvic fins, anal area, and underside of tail, which are prominent dark brown or grey. Aleutian Skate, *Bathyraja aleutica*

57b. Ventral surface, including tail, is white without darker mottling. Sandpaper Skate, *Bathyraja kincaidii*

58a. Disc thorns present. 59.

58b. Disc thorns absent, except for weakly developed nuchal thorns occasionally present; if present less than 3. 62.

59a. Thorns continuous from scapular region to first dorsal fin origin. Commander Skate, *Bathyraja lindbergi*

59b. Thorns discontinuous from scapular region to first dorsal fin origin. 60.

60a. Dorsal disc with 1–5, average 3, well-developed nuchal thorns; dermal denticles on ventral disc and tail surface prickly, giving it a rough to the touch texture. Deepsea Skate, *Bathyraja abyssicola*

60b. Dermal denticles on ventral disc and tail surface minute, giving it a smooth touch texture. 61.

61a. Nuchal thorns always present, well developed; dorsal disc surface purplish in life and with distinct large white blotches. White-blotched Skate, *Bathyraja maculata*

61b. Nuchal thorns usually absent, if present, poorly developed; dorsal disc surface uniformly coloured except for distinct whitish "brows" along inner orbital margins; no distinct large white blotches. Smallthorn Skate, *Bathyraja minispinosa*

62a. Ventral surface coarsely textured, with a very rough to the touch. Pacific White Skate, *Bathyraja spinosissima*

62b. Ventral surface smooth to the touch. 63.

63a. Ventral surface colouration plum brown to black or brown and white. 64.

63b. Ventral surface whitish to creamy coloured with no other distinct markings. 65.

64a. Dorsal surface colouration is uniformly brown above, ventral surface white except for pectoral fins and pelvic region, which are brown. Fine-spined Skate, *Bathyraja microtrachys*

64b. Dorsal and ventral surface colouration is uniformly plum brown to black, except occasional lighter blotches on abdomen. Roughtail Skate, *Bathyraja trachura*

65a. Tail length greater than disc length. Mud Skate, *Bathyraja taranetzi*

65b. Tail length equal to or less than disc length. 66.

66a. Dorsal surface colouration greenish brown with bright yellow blotches. Butterfly Skate, *Bathyraja mariposa*

66b. Dorsal colouration grey to brown with spots and vermiculations. Okhotsk Skate, *Bathyraja violacea*

67a. Scapular thorns present. Broad Skate, *Amblyraja badia*

67b. Scapular thorns absent. 68.

68a. Pelvic fins shallowly notched. Big Skate, *Beringraja binoculata*

68b. Pelvic fins deeply notched. 69.

69a. Snout very short, bluntly pointed. Pacific Starry Skate, *Raja stellulata*

69b. Snout long, acutely pointed. 70.

70a. Snout extremely long, tapering to an acute point; ventral surface mottled light to dark grey. Longnose Skate, *Raja rhina*

70b. Snout moderately long, acutely pointed; ventral surface white to pale tan. California Skate, *Raja inornata*

71a. Caudal fin well developed. Round Stingray, *Urobatis halleri* (Urotrygonidae)

71b. Caudal fin absent. 72.

72a. Head elevated above disc; eyes and spiracles on sides of head. 73.

72b. Head not elevated above disc; eyes and spiracles on top of head. 75.

73a. Disc width nearly twice the length; tail short, slender, less than disc length; stinging spine short. California Butterfly Ray, *Gymnura marmorata* (Gymnuridae)

73b. Disc width only slightly greater than length; tail long, whip-like, greater than disc length; stinging spine long. 74. (Dasyatidae)

74a. Snout pointed; brown to greyish or blackish above, white below. Diamond Stingray, *Hypanus dipterura*

74b. Snout rounded; uniformly dark purple above and below. Pelagic Stingray, *Pteroplatytrygon violacea*

75a. Head without paired horn-like flaps. Bat Ray, *Myliobatis californica* (Myliobatidae)

75b. Head with paired horn-like flaps. 76. (Mobulidae)

76a. Mouth on front of head; teeth on lower jaw only. Manta Ray, *Manta birostris*

76b. Mouth on underside of head; teeth on both jaws. Spinetail Mobula, *Mobula japonica*

REFERENCES

Ainsley, S.M., Ebert, D.A., Cailliet, G.M., 2011. A comparison of reproductive parameters of the Bering skate, *Bathyraja interrupta*, from two Alaskan large marine ecosystems. Mar. Freshw. Res. 62 (6), 557. http://dx.doi.org/10.1071/MF10140.

Ardizzone, D., Cailliet, G.M., Natanson, L.J., Andrews, A.H., Kerr, L.A., Brown, T.A., 2006. Application of bomb radiocarbon chronologies to shortfin mako (Isurus oxyrinchus) age validation. In: Special Issue: Age and Growth of Chondrichthyan Fishes: New Methods, Techniques and Analysis. Springer, The Netherlands, pp. 355–366.

Aschliman, N., Nishida, M., Miya, M., Inoue, J.G., Rosana, K.M., Naylor, G.J.P., 2012. Body plan convergence in the evolution of skates and rays (Chondrichthyes: Batoidea). Mol. Phylogenet. Evol. 63, 28–42.
Baduini, C.L., 1995. Feeding ecology of the basking shark (*Cetorhinus maximus*) relative to distribution and abundances of prey. California State University, Moss Landing Marine LaboratoriesMaster's thesis.
Barker, J., Bartoli, A., Clark, M., Dulvy, N., Gordon, C., Hood, A., Jimenez Alvarado, D., Lawson, J.M., Meyers, E., 2016. Angelshark Action Plan for the Canary Islands. Zoological Society of London Technical Publication. 28 pp.
Barnett, L.A., Winton, M.V., Ainsley, S.M., Cailliet, G.M., Ebert, D.A., 2013. Comparative demography of skates: life-history correlates of productivity and implications for management. PLoS One 8 (5), e65000.
Beebe, M., Tee-Van, J., 1941. Eastern Pacific Expedition of the New York Zoological Society. XXV. Fishes from the tropical Eastern Pacific. Part 2. Sharks. Zoologica 26, 93–122.
Bigman, J.S., Ebert, D.A., Goldman, K.J., 2016. *Squalus suckleyi*. The IUCN Red List of Threatened Species 2016: e.T195488A2382480. Available at: http://dx.doi.org/10.2305/IUCN.UK.2016-1.RLTS.T195488A2382480.en (accessed 24/12/2016).
Boyle, M.D., Ebert, D.A., Cailliet, G.M., 2012. Stable-isotope analysis of a deep-sea benthic-fish assemblage: evidence of an enriched benthic food web. J. Fish Biol. 80 (5), 1485–1507.
Briggs, J.C., 1995. Global Biogeography. Elsevier, Amsterdam.
Cailliet, G.M., 1990. Elasmobranch age determination and verification: an updated review. In: NOAA Technical Report NMFS, 90, pp. 157–165.
Cailliet, G.M., Goldman, K., 2004. Age determination and validation in chondrichthyan fishes. In: Carrier, J.C., Musick, J.A., Heithaus, M.R. (Eds.), Biology of Sharks and Their Relatives. CRC Press, Boca Raton, FL, pp. 399–447.
Cailliet, G.M., Chabot, C.L., Nehmens, M.C., Carlisle, A.B., 2016. *Squatina californica*. The IUCN Red List of Threatened Species 2016: e.T39328A80671059. Available at: http://dx.doi.org/10.2305/IUCN.UK.2016-2.RLTS.T39328A80671059.en (accessed 25/12/2016).
California Department of Fish and Wildlife, 2015. California's Marine Protected Area (MPA) Network. Monterey, CA. Available at: http://www.dfg.ca.gov/marine/mpa/mpa_summary.asp (accessed 27/01/2015).
Campana, S.E., Natanson, L.J., Myklevoll, S., 2002. Bomb dating and age determination of large pelagic sharks. Can. J. Fish. Aquat. Sci. 59, 450–455.
Carlisle, A.B., Perle, C.R., Goldman, K.J., Block, B.A., 2011. Seasonal changes in depth distribution of salmon sharks (*Lamna ditropis*) in Alaskan waters: implications for foraging ecology. Can. J. Fish. Aquat. Sci. 68 (11), 1905–1921.
Carlisle, B., Goldman, K.J., Madigan, D.J., Bigman, J.S., Swithenbank, A.M., Kline, T.C., Block, B.A., 2014. Stable isotope analysis of vertebrae reveals ontogenetic changes in habitat in an endothermic pelagic shark. Proc. R. Soc. B 282 (1799).
Carlisle, A., Litvin, S., Hazen, E., Madigan, D., Goldman, K., Lea, R., Block, B., 2015. Reconstructing habitat use by juvenile salmon sharks links upwelling to strandings in the California Current. Mar. Ecol. Prog. Ser. 525, 217–228.
Carlson, J.K., Parsons, G.R., 1997. Age and growth of the bonnethead shark, *Sphyrna tiburo*, from northwest Florida, with comments on clinal variation. Environ. Biol. Fishes 50 (3), 331–341.
Charnov, E.L., 1993. Life History Invariants. Oxford Series in Ecology and Evolution. Oxford University Press, Oxford. 167 pp.
Compagno, L.J.V., 1973. Interrelationships of living elasmobranchs. In: Greenwood, P.H., Miles, R.S., Patterson, C. (Eds.), Interrelationships of Fishes, pp. 15–61. Zool. J. Linn. Soc. 53 (Supp. 1).
Compagno, L.J.V., 1977. Phyletic relationships of living sharks and rays. Am. Zool. 17 (2), 303–322.

Compagno, L.J.V., 1984. Sharks of the world. An annotated and illustrated catalogue of shark species known to date. In: Part 1. Hexanchiformes to Lamniformes. Part 2. Carcharhiniformes. FAO Species Catalogue for Fishery Purposes. No. 125, Vol. 4. FAO, Rome.

Compagno, L.J.V., 1990. Alternate life history styles of cartilaginous fishes in time and space. Environ. Biol. Fishes 28, 33–75.

Compagno, L.J.V., 2001. Sharks of the world. An annotated and illustrated catalogue of shark species known to date. In: Bullhead, Mackerel, and Carpet Sharks (Heterodontiformes, Lamniformes, and Orectolobiformes). FAO Species Catalogue for Fishery Purposes. No. 1, Vol. 2. FAO, Rome. 269 pp.

Compagno, L.J.V., 2005. Checklist of living Chondrichthyes. In: Hamlett, W.C. (Ed.), Reproductive Biology and Phylogeny of Chondrichthyes: Sharks, Batoids, and Chimaeras. Science Publishers, Inc., Enfield, New Hampshire, pp. 501–548.

Compagno, L.J.V., Ebert, D.A., Smale, M.J., 1989. Guide to the Sharks and Rays of Southern Africa. Struik Publishers, Cape Town. 160 pp.

Cortés, E., 1997. A critical review of methods of studying fish feeding based on analysis of stomach contents: application to elasmobranch fishes. Can. J. Fish. Aquat. Sci. 54 (3), 726–738.

Cortés, E., 1998. Demographic analysis as an aid in shark stock assessment and management. Fish. Res. 39, 199–208.

Cortés, E., 1999. Standardized diet compositions and trophic levels of sharks. ICES J. Mar. Sci. 56, 707–719.

Cortés, E., 2002. Incorporating uncertainty into demographic modeling: application to shark populations and their conservation. Conserv. Biol. 16, 1048–1062.

Cortés, E., 2004. Life history patterns, demography, and population dynamics. In: Carrier, J., Musick, J., Heithaus, M.R. (Eds.), Biology of Sharks and Their Relatives. CRC Press, New York, pp. 449–470.

Cortés, E., Brooks, E.N., Gedamke, T., 2012. Population dynamics, demography, and stock assessment. In: Biology of Sharks and Their Relatives. Boca Raton, FL, pp. 453–486.

COSEWIC, 2007. COSEWIC Status and Assessment Report on the Basking Shark *Cetorhinus maximus* (Pacific Population) in Canada. Committee on the Status of Endangered Wildlife in Canada, Ottawa, Canada. Available at: http://www.sararegistry.gc.ca/status/status_e.cfm (accessed 25/12/2016).

Cox, G., Francis, M., 1997. Sharks and Rays of New Zealand. Canterbury University Press, University of Canterbury, Christchurch, New Zealand, 68 pp.

Davis, C.D., Cailliet, G.M., Ebert, D.A., 2007. Age and growth of the roughtail skate *Bathyraja trachura* (Gilbert 1892) from the eastern North Pacific. Environ. Biol. Fishes 80, 325–336.

de Carvalho, M.R., 1996. Higher-level elasmobranch phylogeny, basal squaleans, and paraphyly. In: Stiassny, M.L.J., Parenti, L.R., Johnson, G.D. (Eds.), Interrelationships of Fishes. Academic Press, San Diego, London, pp. 35–62.

Denney, N.H., Jennings, S., Reynolds, J.D., 2002. Life-history correlates of maximum population growth rates in marine fishes. Proc. Biol. Sci. 269 (1506), 2229–2237. http://dx.doi.org/10.1098/rspb.2002.2138.

Department of Fisheries and Oceans Canada, 2014. Canada–British Columbia Marine Protected Area Network Strategy. Her Majesty the Queen in Right of Canada, 35 pp.

Didier, D.A., Kemper, J.M., Ebert, D.A., 2012. Phylogeny, biology, and classification of extant Holocephali. In: Carrier, J.C., Musick, J.A., Heithaus, M.R. (Eds.), TheBiology of Sharks and Their Relatives, vol. 1. CRC Press, Boca Raton, FL, pp. 97–121 (Chapter 4).

Douady, C.J., Dosay, M., Shivji, M.S., Stanhope, M.J., 2003. Molecular phylogenetic evidence refuting the hypothesis of Batoidea (rays and skates) as derived sharks. Mol. Phylogenet. Evol. 26, 215–221.

Dulvy, N.K., Forrest, R.E., 2010. Life histories, population dynamics, and extinction risks in chondrichthyans. In: Sharks and Their Relatives, vol. 2. pp. 639–679.

Dulvy, N.K., Sadovy, Y., Reynolds, J.D., 2003. Extinction vulnerability in marine populations. Fish Fish. 4 (1), 25–64.

Dulvy, N.K., Ellis, J.R., Goodwin, N.B., Grant, A., Reynolds, J.D., Jennings, S., 2004. Methods of assessing extinction risk in marine fishes. Fish Fish. 5 (3), 255–276.

Dulvy, N.K., Baum, J.K., Clarke, S., Compagno, L.J.V., Cortès, E., Domingo, A., Fordham, S., Fowler, S., Francis, M.P., Gibson, C., Martìnez, J., Musick, J.A., Soldo, A., Stevens, J.D., Valenti, S., 2008. You can swim but you can't hide: global status and conservation of oceanic pelagic sharks and rays. Aquat. Conserv. Mar. Freshw. Ecosyst. 44, 303–313.

Dulvy, N.K., Fowler, S.L., Musick, J.A., Cavanagh, R.D., Kyne, P.M., Harrison, L.R., Carlson, J.K., Davidson, L.N.K., Fordham, S.V., Francis, M.P., Pollock, C.M., Simpfendorfer, C.A., Burgess, G.H., Carpenter, K.E., Compagno, L.J.V., Ebert, D.A., Gibson, C., Heupel, M.R., Livingstone, S.R., Sanciangco, J.C., Stevens, J.D., Valenti, S., White, W.T., 2014a. Global extinction risk of sharks and rays: threats, hotspots, and conservation. eLife 3, e00590. http://dx.doi.org/10.7554/eLife.00590.

Dulvy, N.K., Pardo, S.A., Simpfendorfer, C.A., Carlson, J.K., 2014b. Diagnosing the dangerous demography of manta rays using life history theory. Peer J. 2, e400.

Dulvy, N.K., Allen, D.J., Ralph, G.M., Walls, R.H.L., 2016. The Conservation Status of Sharks, Rays, and Chimaeras in the Mediterranean Sea. IUCN Centre for Mediterranean Cooperation, Campanillas, Málaga, Spain. 14 pp.

Ebert, D.A., 2001. First eastern Pacific records of the longfin mako shark, *Isurus paucus* Guitar-Manday, 1966. Calif. Fish Game 87 (3), 117–121.

Ebert, D.A., 2003. The Sharks, Rays and Chimaeras of California. University of California Press, Los Angeles, CA, 284 pp.

Ebert, D.A., 2005. Reproductive biology of skates, *Bathyraja* (Ishiyama), along the eastern Bering Sea continental slope. J. Fish Biol. 66, 618–649.

Ebert, D.A., 2016. Deep-sea cartilaginous fishes of the southeastern Pacific Ocean. In: FAO Species Catalogue for Fishery Purposes. FAO, Rome, Italy. 241 pp.

Ebert, D.A., Bizzarro, J.J., 2007. Standardized diet compositions and trophic levels of skates (Chondrichthyes: Rajiformes: Rajoidei). Environ. Biol. Fishes 80 (2–3), 221–237.

Ebert, D.A., van Hees, K.E., 2015. Beyond jaws: rediscovering the "*Lost Sharks*" of southern Africa. In: Ebert, D.A., Huveneers, C., Dudley, S.F.J. (Eds.), Advances in Shark Research, pp. 141–156. Afr. J. Mar. Sci. 37.

Ebert, D.A., Winton, M.V., 2010. Chondrichthyans of high latitude seas. In: Carrier, J.C., Musick, J.A., Heithaus, M.R. (Eds.), TheBiology of Sharks and Their Relatives, vol. 2. CRC Press, Boca Raton, FL, pp. 115–158.

Ebert, D.A., Mollet, H.F., Baldridge, A., Thomas, T., Forney, K., Ripley, W.E., 2004. Occurrence of the whale shark, *Rhincodon typus* Smith 1828, in California waters. Northwest. Nat. 85 (1), 26–28.

Ebert, D.A., White, W.T., Goldman, K.J., Compagno, L.J.V., Daly-Engel, T.S., Ward, R., 2010. Reevaluation and redescription of *Squalus suckleyi* (Girard, 1854) from the North Pacific, with comments on the *Squalus acanthias* subgroup (Squaliformes: Squalidae). Zootaxa 2612, 22–40.

Ebert, D.A., Ho, H.-C., White, W.T., de Carvalho, M.R., 2013a. Introduction to the systematics and biodiversity of sharks, rays, and chimaeras (Chondrichthyes) of Taiwan. In: Carvalho, M.R., Ebert, D.A., Ho, H.-C., White, W.T. (Eds.), Systematics and Biodiversity of Sharks, Rays, and Chimaeras (Chondrichthyes) of Taiwan, pp. 005–019. Zootaxa 3752 (1).

Ebert, D.A., Fowler, S., Compagno, L.J.V., 2013b. Sharks of the World: A Fully Illustrated Guide to the Sharks of the World. Wild Nature Press, Plymouth, UK. 528 pp.

Ebert, D.A., Knuckey, J.D.S., Kamikawa, D.J., 2014. First eastern North Pacific record of the velvet dogfish, *Zameus squamulosus* (Chondrichthyes: Squaliformes: Somniosidae). Mar. Biodivers. Rec. 7, e48, 1–3. http://dx.doi.org/10.1017/S1755267214000499.

Ebert, D.A., Pien, C., Kamikawa, D.J., 2015. Confirmation of the cookiecutter shark, *Isistius brasiliensis*, from the eastern North Pacific Ocean (Squaliformes: Dalatiidae). Mar. Biodivers. Rec. 8, e1183.

Eckert, S.A., Stewart, B.S., 2001. Telemetry and satellite tracking of whale sharks, *Rhincodon typus*, in the Sea of Cortez, Mexico, and the north Pacific Ocean. Environ. Biol. Fishes 60 (1–3), 299–308.

Ellis, J., Soldo, A., Dureuil, M., Fordham, S., 2015. *Squalus acanthias*. The IUCN Red List of Threatened Species 2015: e.T91209505A48910866 (accessed 24/12/2016).

Eschmeyer, W.N., Fricke, R., van der Laan, R., 2016. Catalog of Fishes: Genera, Species. http://researcharchive.calacademy.org/research/ichthyology/catalog/fishcatmain.asp (accessed 23/12/2016).

Estrada, J.A., Rice, A.N., Lutcavage, M.E., Skomal, G.B., 2003. Predicting trophic position in sharks of the north-west Atlantic Ocean using stable isotope analysis. J. Mar. Biol. Assoc. U. K. 83, 1347–1350.

Estrada, J.A., Rice, A.N., Natanson, L.J., Skomal, G.B., 2006. Use of isotopic analysis of vertebrae in reconstructing ontogenetic feeding ecology in white sharks. Ecology 87 (4), 829–834.

Farrugia, T.J., Goldman, K.J., King, J.R., Ormseth, O.A., 2016. *Beringraja binoculata*. The IUCN Red List of Threatened Species 2016: e.T44183A80679344. Available at: http://dx.doi.org/10.2305/IUCN.UK.2016-2.RLTS.T44183A80679344.en (accessed 24/12/2016).

Fisk, A.T., Tittlemier, S.A., Pranschke, J.L., Norstrom, R.J., 2002. Using anthropogenic contaminants and stable isotopes to assess the feeding ecology of Greenland sharks. Ecology 83 (8), 2162–2172.

Fordham, S., Fowler, S.L., Coelho, R.P., Goldman, K.J., Francis, M.P., 2016. *Squalus acanthias*. The IUCN Red List of Threatened Species 2016: e.T91209505A2898271. Available at: http://dx.doi.org/10.2305/IUCN.UK.2016-1.RLTS.T91209505A2898271.en (accessed 24/12/2016).

Fowler, S.L., 2009. *Cetorhinus maximus* (North Pacific Subpopulation). The IUCN Red List of Threatened Species 2009: e.T39545A10246972. Available at: http://dx.doi.org/10.2305/IUCN.UK.2009-2.RLTS.T39545A10246972.en (accessed 24/12/2016).

Frisk, M.G., Miller, T.J., Dulvy, N.K., 2005. Life histories and vulnerability to exploitation of elasmobranchs: inferences from elasticity, perturbation and phylogenetic analyses. J. Northwest Atl. Fish. Sci. 35, 27–45.

Gallucci V., Taylor I., King J., McFarlane G. and McPhie R., Spiny Dogfish (*Squalus acanthias*) Assessment and Catch Recommendations for 2010, 2011, Department of Fisheries and Oceans Canada Canadian Science Advisory Secretariat Research Document 034, 71 pp.

García, V.B., Lucifora, L.O., Myers, R.A., 2008. The importance of habitat and life history to extinction risk in sharks, skates, rays and chimaeras. Proc. R. Soc. B 275, 83–89.

Gburski, C.M., Gaichas, S.K., Kimura, D.K., 2007. Age and growth of the big skate (*Raja binoculata*) and longnose skate (R. rhina) in the Gulf of Alaska. Environ. Biol. Fishes 80, 337–349.

Gertseva, V.V., Schirripa, M.J., 2008. Status of the Longnose Skate (*Raja rhina*) off the Continental US Pacific Coast in 2007. Pacific Fishery Management Council, Newport, OR. 131 pp.

Goldman, K.J., 2005. 6. Age and growth of elasmobranch fishes. In: Management Techniques for Elasmobranch Fisheries, vol. 474. pp. 76.

Grubbs, R. D., 2010. Ontogenetic Shifts in Movements and Habitat Use. Sharks and Their Relatives II: Biodiversity, Adaptive Physiology, and Conservation. CRC Press, Boca Raton, FL, pp. 319–350.
Gunderson, D.R., 1980. Using rK selection theory to predict natural mortality. Can. J. Fish. Aquat. Sci. 37 (12), 2266–2271.
Hart, J.L., 1973. Pacific fishes of Canada. Bull. Fish. Res. Board Can. 180, 740 pp.
Heithaus, M.R., Frid, A., Vaudo, J.J., Worm, B., Wirsing, A.J., 2010. Unraveling the ecological importance of elasmobranchs. In: Sharks and Their Relatives II: Biodiversity, Adaptive Physiology, and Conservation. CRC Press, Boca Raton, FL, pp. 611–637.
Hoenig, J.M., Gruber, S.H., 1990. Life-history patterns in the elasmobranchs: implications for fisheries management. NOAA Tech. Rep. NMFS 90 (1), 16.
Holden, M.J., Rout, D.W., Humphreys, C.N., 1971. The rate of egg laying by three species of ray. J. Cons. Int. Explor. Mer. 33 (3), 335–339.
Hussey, N.E., Dudley, S.F., McCarthy, I.D., Cliff, G., Fisk, A.T., 2011. Stable isotope profiles of large marine predators: viable indicators of trophic position, diet, and movement in sharks? Can. J. Fish. Aquat. Sci. 68 (12), 2029–2045.
Hutchings, J.A., Myers, R., Garcia, V.B., Lucifora, L.O., Kuparinen, A., 2012. Life-history correlates of extinction risk and recovery potential. Ecol. Appl. 23 (3), 515–522.
Hyslop, E.J., 1980. Stomach contents analysis—a review of methods and their application. J. Fish Biol. 17 (4), 411–429.
Ishihara, H., Mochizuki, T., Homma, K., Taniuchi, T., 2002. Reproductive strategy of the Japanese common skate (Spiny rasp skate) *Okameji kenojei*. In: Fowler, S.L., Reed, T.M., Dipper, F.A. (Eds.), Elasmobranch Biodiversity, Conservation and Management: Proceedings of the International Seminar and Workshop, Sabah, Malaysia, July 1997. IUCN SSC Shark Specialist Group. IUCN, Gland, Switzerland and Cambridge, pp. 236–240.
IUCN, 2012. Guidelines for Application of IUCN Red List Criteria at Regional and National Levels: Version 4.0. IUCN, Gland, Switzerland and Cambridge, UK: iii + 41 pp.
James, K.C., Ebert, D.A., Long, D.J., Didier, D.A., 2009. A new species of chimera, *Hydrolagus melanophasma* sp. nov. (Chondrichthyes: Chimaeriformes: Chimaeridae), from the eastern North Pacific. Zootaxa 2218, 59–68.
James, K.C., Ebert, D.A., Natanson, L.J., Cailliet, G.M., 2014. Age and growth characteristics of the Starry Skate, *Raja stellulata*, with a description of life history and habitat trends of the central California, USA, skate assemblage. Environ. Biol. Fishes 97 (4), 435–448.
Jennings, S., Reynolds, J.D., Mills, S.C., 1998. Life history correlates of responses to fisheries exploitation. Proc. R. Soc. Lond. B Biol. Sci. 265 (1393), 333–339.
Jennings, S., Reynolds, J.D., Polunin, N.V., 1999. Predicting the vulnerability of tropical reef fishes to exploitation with phylogenies and life histories. Conserv. Biol. 13 (6), 1466–1475.
Jones, P.J.S., Qiu, W., De Santo, E.M., 2013. Governing marine protected areas: social–ecological resilience through institutional diversity. Mar. Policy 41, 5–13.
Joung, S.J., Chen, C.T., Clark, E., Uchida, S., Huang, W.Y., 1996. The whale shark, *Rhincodon typus*, is a livebearer: 300 embryos found in one 'megamamma' supreme. Environ. Biol. Fishes 46 (3), 219–223.
Juan-Jordá, M.J., Mosqueira, I., Freire, J., Dulvy, N.K., 2013. Life history correlates of marine fisheries vulnerability: a review and a test with tunas and mackerel species. In: Briand, F. (Ed.), CIESM Workshop Monograph No. 45. CIESM Publisher, Monaco, pp. 113–128.
Juan-Jordá, M.J., Mosqueira, I., Freire, J., Dulvy, N.K., 2015. Population declines of tuna and relatives depend on their speed of life. Proc. R. Soc. Lond. B Biol. Sci. 282 (1811), 20150322.

Ketchen, K.S., 1975. Age and growth of dogfish *Squalus acanthias* in British Columbia waters. J. Fish. Board Can. 32 (1), 43–59.
Ketchen, K.S., 1986. The spiny dogfish (*Squalus acanthias*) in the Northeast Pacific and a history of its utilization. Can. Spec. Publ. Fish. Aquat. Sci. 88, 78.
Kim, S.L., Tinker, M.T., Estes, J.A., Koch, P.L., 2012. Ontogenetic and among-individual variation in foraging strategies of northeast Pacific white sharks based on stable isotope analysis. PLoS One 7 (9), e45068.
Kyne, P.M., Simpfendorfer, C.A., 2010. Chondrichthyans of high latitude seas. In: Carrier, J.C., Musick, J.A., Heithaus, M.R. (Eds.), TheBiology of Sharks and Their Relatives, vol. 2. CRC Press, Boca Raton, FL, pp. 37–113 (Chapter 2).
Last, P.R., White, W.T., de Carvalho, M.R., Séret, B., Stehmann, M.F.W., Naylor, G.J.P., 2016. Rays of the World. Cornell University Press, Ithaca, NY. 832 pp.
Leet, W.S., Dewees, C.M., Klingbeil, R., Larson, E.J., 2001. California's Living Marine Resources: A Status Report. The Resources Agency, California Department of Fish and Game. 585 pp.
Licandeo, R., Cerna, F.T., 2007. Geographic variation in life-history traits of the endemic kite skate *Dipturus chilensis* (Batoidea: Rajidae), along its distribution in the fjords and channels of southern Chile. J. Fish Biol. 71 (2), 421–440.
Liu, K.M., Chen, C.T., 1999. Demographic analysis of the scalloped hammerhead, *Sphyrna lewini*, in the northwestern Pacific. Fish. Sci. 65 (2), 218–223.
Lombardi-Carlson, L.A., Cortés, E., Parsons, G.R., Manire, C.A., 2003. Latitudinal variation in life-history traits of bonnethead sharks, *Sphyrna tiburo*, (Carcharhiniformes: Sphyrnidae) from the eastern Gulf of Mexico. Mar. Freshw. Res. 54, 875–883.
Maurer, J.R.F., 2009. Life history of two Bering Sea slope skates: *Bathyraja lindbergi* and *B. maculata*. Master's thesisCalifornia State University, Moss Landing Marine Laboratories.
McClenachan, L., Cooper, A.B., Carpenter, K.E., Dulvy, N.K., 2012. Extinction risk and bottlenecks in the conservation of charismatic marine species. Conserv. Lett. 5 (1), 73–80.
McFarlane, G.A., Beamish, R.J., 1987. Validation of the dorsal spine method of age determination for spiny dogfish. In: Age and Growth of Fish. Iowa State University Press, Ames, IA, pp. 287–300.
McFarlane, G.A., King, J.R., 2006. Age and growth of big skate (*Raja binoculata*) and longnose skate (*Raja rhina*) in British Columbia waters. Fish. Res. 78, 169–178.
McFarlane, G.A., King, J.R., Saunders, M.W., 2002. Preliminary study on the use of neural arches in the age determination of bluntnose sixgill sharks (*Hexanchus griseus*). Fish. Bull. 100 (4), 861–864.
McFarlane, G.A., McPhie, R.P., King, J.R., 2010. Distribution and Life History Parameters of Elasmobranch Species in British Columbia Waters. In: Canadian Technical Report of Fisheries and Aquatic Sciences 2908. 157 pp.
Mecklenburg, C.W., Mecklenburg, T.A., Thorsteinson, L.K., 2002. Fishes of Alaska. American Fisheries Society, Bethesda, MD. 1037 pp.
Morris, D.W., 2003. Toward an ecological synthesis: a case for habitat selection. Oecologia 136 (1), 1–13.
Mull, C.G., Lowe, C.G., Young, K.A., 2010. Seasonal reproduction of female round stingrays (*Urobatis halleri*): steroid hormone profiles and assessing reproductive state. Gen. Comp. Endocrinol. 166 (2), 379–387.
Musick, J.A., Ellis, J.K., 2005. Reproductive evolution of chondrichthyans. In: Hamlett, W.C., Jamieson, B.G.M. (Eds.), Reproductive Biology and Phylogeny of Chondrichthyes: Sharks, Batoids, and Chimaeras. Science Publishers, Enfield, NH, pp. 45–47.
Myers, R.A., Mertz, G., Fowlow, P.S., 1997. Maximum population growth rates and recovery times for Atlantic cod. Fish. Bull. 95 (762), 772.

Myers, R.A., Bowen, K.G., Barrowman, N.J., 1999. Maximum reproductive rate of fish at low population sizes. Can. J. Fish. Aquat. Sci. 56 (12), 2404–2419.

Natanson, L.J., Cailliet, G.M., 1990. Vertebral growth zone deposition in Pacific angel sharks. Copeia 1990 (4), 1133–1145.

Natanson, L.J., Casey, J.G., Kohler, N.E., 1995. Age and growth estimates for the dusky shark, *Carcharhinus obscurus*, in the western North Atlantic Ocean. Oceanogr. Lit. Rev. 9 (42), 786.

Naylor, G.J.P., Ryburn, J.A., Fedrigo, O., Lopez, A., 2005. Phylogenetic relationships among the major lineages of modern elasmobranchs. In: Hamlett, W.C. (Ed.), Reproductive Biology and Phylogeny of Chondrichthyes—Sharks, *Batoids and Chimaeras*. Science Publisher, Inc., Enfield, NH, USA, pp. 1–25.

Naylor, G.J.P., Caira, J.N., Jensen, K., Rosana, K.A.M., White, W.T., Last, P.R., 2012a. Bull. Am. Nat. Hist. Mus. 367, 1–263ADNA sequence-based approach to the identification of shark and ray species and its implications for global elasmobranch diversity and parasitology,

Naylor, G.J.P., Caira, J.N., Jensen, K., Kerri, Rosana, K.A.M., Straube, N., Lakner, C., 2012b. Elasmobranch phylogeny: a mitochondrial estimate based on 595 species. In: Carrier, J.C., Musick, J.A., Heithaus, M.R. (Eds.), Biology of Sharks and Their Relatives, second ed. CRC Press, Boca Raton, London, New York, pp. 31–56.

Neer, J.A., Thompson, B.A., 2005. Life history of the cownose ray, Rhinoptera bonasus, in the northern Gulf of Mexico, with comments on geographic variability in life history traits. Environ. Biol. Fishes 73 (3), 321–331.

Nelson, J.S., Grande, T.C., Wilson, M.V.H., 2016. Fishes of the World, fifth ed. John Wiley & Sons, Hoboken, NJ, p. 707.

Notarbartolo-di-Sciara, G., 1988. Natural history of the rays of the genus *Mobula* in the Gulf of California. Fish. Bull. 86 (1), 45–66.

Ormseth, O.A., 2014a. Assessment of the skate stock complex in the Bering Sea and Aleutian Islands. In: Stock Assessment and Fishery Evaluation Report for the Groundfish Resources of the Bering Sea and Aleutian Islands Region. National Marine Fisheries Service Alaska Fisheries Science Center, Seattle, WA, pp. 1693–1782.

Ormseth, O.A., 2014b. Assessment of the skate stock complex in the Gulf of Alaska. In: Stock Assessment and Fishery Evaluation Report for the Groundfish Resources of the Gulf of Alaska Region. National Marine Fisheries Service Alaska Fisheries Science Center, Seattle, WA, pp. 855–920.

Orr, J.W., Stevenson, D.E., Hoff, G.R., Spies, I., McEchron, J.D., 2011. *Bathyraja panthera*, a new species of skate (Rajidae: Arhynchobatinae) from the western Aleautian Islands, and resurrection of the subgenus *Arctorja* Ishiyama. NOAA Prof. Pap. NMFS. 1150 pp.

Pardo, S.A., Kindsvater, H.K., Cuevas-Zimbrón, E., Sosa-Nishizaki, O., Pérez-Jiménez, J.C., Dulvy, N.K., 2016. Growth, Productivity, and Relative Extinction Risk of a Data-Sparse Devil Ray. Sci. Rep. 6, 33745.

Pauly, D., 1980. On the interrelationships between natural mortality, growth parameters, and mean environmental temperature in 175 fish stocks. ICES J. Mar. Sci. 39 (2), 175–192.

Pauly, D., 2002. Growth and mortality of the basking shark Cetorhinus maximus and their implications for management of whale sharks Rhincodon typus. In: Fowler, S.L., Reed, T.M., Dipper, F.A. (Eds.), Elasmobranch Biodiversity, Conservation and Management. Proceedings of the International Seminar and Workshop, Sabah, Malaysia, July 1997. IUCN SSC Shark Specialist Group. IUCN, Gland, Switzerland and Cambridge, UK, pp. 199–208.

Peres, M.B., Vooren, C.M., 1991. Sexual development, reproductive cycle, and fecundity of the school shark *Galeorhinus galeus* off southern Brazil. Fish. Bull. 89 (4), 655–667.

Peterson, B.J., Fry, B., 1987. Stable isotopes in ecosystem studies. Annu. Rev. Ecol. Syst. 18, 293–320.

Pianka, E.R., 1970. On r- and K-selection. Am. Nat. 104 (940), 592–597.
Reichert, A.N., Lunsten, L., Ebert, D.A., 2016. First North Pacific records of the pointy nosed blue chimaera, *Hydrolagus* cf. *trolli* (Chondrichthyes: Chimaeriformes: Chimaeridae). Mar. Biodivers. Rec. 9, 90.
Reynolds, J.D., Jennings, S., Dulvy, N.K., 2001. Life histories of fishes and population responses to exploitation. In: Reynolds, J.D., Mace, G.M., Redford, K.H., Robinson, J.G. (Eds.), Conservation of Exploited Species. Cambridge University Press, Cambridge, pp. 148–168.
Ribot-Carballal, M.C., Galvan-Magaña, F., Quiñonez-Velazquez, C., 2005. Age and growth of the shortfin mako shark, *Isurus oxyrinchus*, from the western coast of Baja California Sur, Mexico. Fish. Res. 76 (1), 14–21.
Roff, D.A., 2002. Life History Evolution. Sinauer Associates, Sunderland, Massachusetts. 527 pp.
Shirai, S., 1992. Squalean Phylogeny. A New Framework of "Squaloid" Sharks and Related Taxa. Hokkaido U. Press, Sapporo, pp. 1–151.
Shirai, S., 1996. Phylogenetic interrelationships of neoselachians (Chondrichthyes: Euselachii). In: Stiassny, M.L.J., Parenti, L.R., Johnson, G.D. (Eds.), Interrelationships of Fishes. Academic Press, San Diego, London, pp. 9–34.
Simpfendorfer, C.A., 2005. Demographic models: life tables, matrix models and rebound potential. In: FAO Fisheries Technical Paper, 474. 143.
Smith, W.D., 2004. Life history aspects and population dynamics of a commercially exploited stingray, *Dasyatis Dipterura*. M.Sc. thesis. Moss Landing Marine Laboratories, California, USA.
Stearns, S.C., 1983. The genetic basis of differences in life-history traits among six populations of mosquitofish (*Gambusia affinis*) that shared ancestors in 1905. Evolution 37 (3), 618–627.
Stearns, S.C., 1992. The Evolution of Life Histories. vol. 249. Oxford University Press, Oxford.
Stevens, J.D., 2010. Epipelagic oceanic elasmobranchs. In: Carrier, J.C., Musick, J.A., Heithaus, M.R. (Eds.), Sharks and Their Relatives II. CRC Press, Boca Raton, FL, pp. 3–35 (Chapter 1).
Stevenson, D., 2004. Identification of skates, sculpins, and smelts by observers in north Pacific groundfish fisheries (2002-2003). In: U.S. Department of Commerce Technical Memorandum NMFS-AFSC-142. 67 pp.
Stevenson, D.E., Orr, J.W., Hoff, G.R., McEachran, J.D., 2004. *Bathyraja mariposa*: a new species of skate (Rajidae: Arhynchobatinae) from the Aleutian Islands. Copeia 2004 (2), 305–314.
Stevenson, D.E., Orr, J.W., Hoff, G.R., McEachran, J.D., 2008. Emerging patterns of species richness, diversity, population density, and distribution in the skates (Rajidae) of Alaska. Fish. Bull. 106 (1), 24–39.
Tanaka, S., 1990. Age and Growth Studies on the Calcified Structures of Newborn Sharks in Laboratory Aquaria Using Tetracycline. In: Elasmobranchs as living resources: advances in the biology, ecology, systematics, and the status of the fisheries. National Oceanic and Atmospheric Administration, Technical Report, National Marine Fisheries Service, 90, pp. 189–202.
Tribuzio, C.A., Kruse, G.H., 2012. Life history characteristics of a lightly exploited stock of *Squalus suckleyi*. J. Fish Biol. 80 (5), 1159–1180.
Tribuzio, C.A., Echave, K., Rodgveller, C., Hulson, P., Goldman, K.J., 2011. Assessment of the shark stock complex in the Gulf of Alaska. In: Stock Assessment and Fishery Evaluation Report for the Groundfish Resources of the Gulf of Alaska. North Pacific Fishery Management Council, Anchorage, AK, pp. 1393–1446.
United Nations, 2015. Transforming our world: the 2030 Agenda for Sustainable Development. In: United Nations. A/RES/70/1. 41 pp.

Van Cleve, F.B., Bargmann, G., Culver, M., the MPA Work Group, 2009. Marine Protected Areas in Washington: Recommendations of the Marine Protected Areas Work Group to the Washington State Legislature. Washington Department of Fish and Wildlife, Olympia, WA. 34 pp + 5 Appendices.

Vaudo, J.J., Heithaus, M.R., 2011. Dietary niche overlap in a nearshore elasmobranch mesopredator community. Mar. Ecol. Prog. Ser. 425, 247–260.

von Bertalanffy, L., 1938. A quantitative theory of organic growth (inquiries on growth laws. II). Hum. Biol. 10 (2), 181–213.

Walter, J.P., Ebert, D.A., 1991. Preliminary estimates of age of the bronze whaler *Carcharhinus brachyurus* (Chondrichthyes: Carcharhinidae) from southern Africa, with a review of some life history parameters. S. Afr. J. Mar. Sci. 10 (1), 37–44.

Wetherbee, B.M., Cortés, E., Bizzarro, J.J., 2012. Food consumption and feeding habits. In: Carrier, J.C., Musick, J.A., Heithaus, M.R. (Eds.), Biology of Sharks and Their Relatives CRC Press, Boca Raton, FL, pp. 239–264.

White, W.T., Sommerville, E., 2010. Elasmobranchs of tropical marine ecosystems. In: Carrier, J.C., Musick, J.A., Heithaus, M.R. (Eds.), Sharks and Their Relatives II. CRC Press, Boca Raton, FL, pp. 159–240 (Chapter 1).

Worm, B., Hilborn, R., Baum, J.K., Branch, T.A., Collie, J.S., Costello, C., Fogarty, M.J., Fulton, E.A., Hutchings, J.A., Jennings, S., Jensen, O.P., Lotze, H.K., Mace, P.M., McClanahan, T.R., Minto, C., Palumbi, S.R., Parma, A.M., Ricard, D., Rosenberg, A.A., Watson, R., Zeller, D., 2009. Rebuilding global fisheries. Science 325 (5940), 578–585.

Wourms, J.P., 1977. Reproduction and development in chondrichthyan fishes. Am. Zool. 17 (2), 379–410.

Wourms, J.P., Demski, L.S., 1993. The reproduction and development of sharks, skates, rays and ratfishes: introduction, history, overview, and future prospects. Environ. Biol. Fishes 38 (1–3), 7–21.

Yudin, K.G., Cailliet, G.M., 1990. Age and growth of the gray smoothhound *Mustelus californicus* and the brown smoothhound, *M. henlei*, sharks from central California. Copeia 1990 (1), 191–204.

Zeiner, S.J., 1991. Growth characteristics and estimates of age at maturity of two species of skates (*Raja binoculata* and *Raja rhina*) from Monterey Bay, California. Master's thesisCalifornia State University, Moss Landing Marine Laboratories.

CHAPTER THREE

Review of Current Conservation Genetic Analyses of Northeast Pacific Sharks

Shawn E. Larson*,[1], Toby S. Daly-Engel†, Nicole M. Phillips‡

*Seattle Aquarium, Seattle, WA, United States
†Florida Institute of Technology, Melbourne, FL, United States
‡The University of Southern Mississippi, Hattiesburg, MS, United States
[1]Corresponding author: e-mail address: s.larson@seattleaquarium.org

"Adult Sixgill Shark *Hexanchus griseus*" with photo credit to Dean Grubbs.

Contents

Advances in Marine Biology, Volume 77
ISSN 0065-2881
http://dx.doi.org/10.1016/bs.amb.2017.06.005

Abstract

Conservation genetics is an applied science that utilizes molecular tools to help solve problems in species conservation and management. It is an interdisciplinary specialty in which scientists apply the study of genetics in conjunction with traditional ecological fieldwork and other techniques to explore molecular variation, population boundaries, and evolutionary relationships with the goal of enabling resource managers to better protect biodiversity and identify unique populations. Several shark species in the northeast Pacific (NEP) have been studied using conservation genetics techniques, which are discussed here. The primary methods employed to study population genetics of sharks have historically been nuclear microsatellites and mitochondrial (mt) DNA. These markers have been used to assess genetic diversity, mating systems, parentage, relatedness, and genetically distinct populations to inform management decisions. Novel approaches in conservation genetics, including next-generation DNA and RNA sequencing, environmental DNA (eDNA), and epigenetics are just beginning to be applied to elasmobranch evolution, physiology, and ecology. Here, we review the methods and results of past studies, explore future directions for shark conservation genetics, and discuss the implications of molecular research and techniques for the long-term management of shark populations in the NEP.

1. BACKGROUND

Conservation genetics is an applied science that utilizes molecular tools to help solve problems in species conservation and management (Frankham, 2010; Hedrick and Miller, 1992). Conservation genetics, in practice, is a large field that encompasses the fields of population genetics, molecular ecology, phylogenetics, phylogeography, and taxonomy, where the immediate goals of studies can vary widely (Ogden et al., 2009). However, most studies have an ultimate long-term goal of preventing the extinction or extirpation of sensitive, endangered, or threatened species, as well as to preserve enough genetic variation so that future adaptation, successful expansion, or reestablishment of the natural population is possible (Hedrick and Miller, 1992). As a result, most studies seek to answer basic questions related to the identification of species and resolution of taxonomy (Frankham, 2010), as well as assess population structure, levels of genetic diversity, relatedness, changes in population size, and effective population size (Primmer, 2009). Such studies generate data that may be unattainable by other means and are ideally used by fish and wildlife managers to develop conservation plans. Despite the valuable information gleaned from conservation genetic studies, they can have some drawbacks, such as the need for large sample sizes for robust results, and the considerable training of personnel required to appropriately analyse and interpret the data. Done rigorously, however, the conservation and economic benefits of this type of research can be considerable (Morin et al., 2010).

In the last quarter of the 20th century, conservation genetics studies primarily focused on captive populations. However, over the past 2 decades, molecular studies on at-risk species in the wild have become more common (Frankham, 2010). Studies on marine species initially focused on economically important populations (Hooker and Gerber, 2004; Nielsen et al., 2009), although in recent years there has been increased attention on threatened or endangered marine vertebrates, including elasmobranchs. There are approximately 500 known species of sharks, over 75% of which are listed as endangered, threatened, near-threatened, or data deficient by the International Union for Conservation of Nature (IUCN), with fewer than 25% listed as Least Concern (Dulvy et al., 2014). Sharks are at high risk of extinction due to habitat loss/degradation, exploitation for their fins and meat, and because they are common fisheries bycatch (see Chapter "Age and growth of elasmobranchs and applications to fisheries management and conservation in the Northeast Pacific Ocean" by Matta et al., this volume). They are especially vulnerable to overexploitation because they are typically 'K selected' species, with life history strategies characterized by late maturity, long lifespans, and low fecundity (Baum and Myers, 2004; Baum et al., 2003; Musick et al., 2000). They have also been found to have low genetic diversity compared with other taxa (Daly-Engel et al., 2010; Martin et al., 1992), making molecular methods increasingly important for addressing conservation concerns in sharks (Fig. 1).

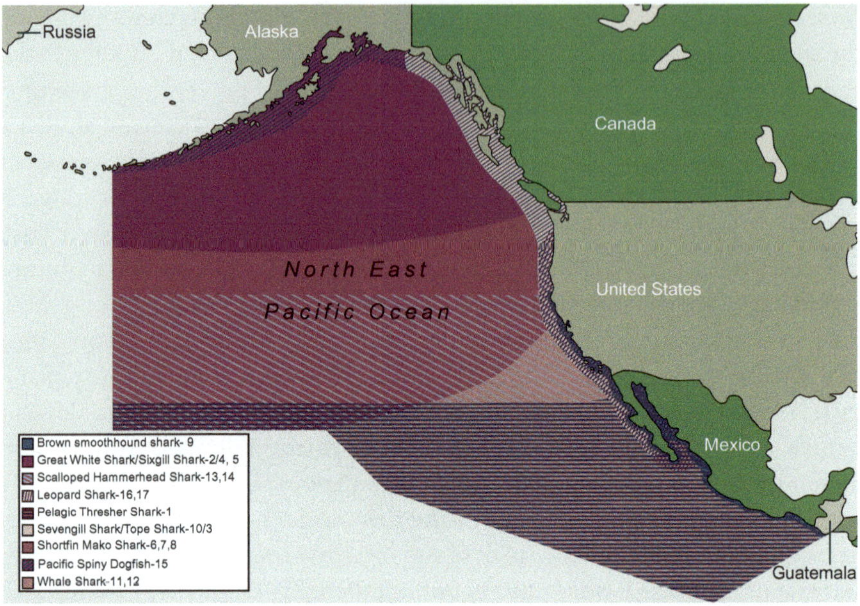

Fig. 1 Map of the northeast Pacific (NEP) with ranges of sharks studied genetically and associated studies indicated by number in the legend. To see studies go to Table 1.

Effective management of many sharks is challenging because of the elusive nature of most species and, as a result, little is often known about their basic demographics and patterns of habitat use. Most sharks found in the northeast Pacific (NEP) are high-level predators occupying important ecological niches that make them critical for ecosystem stability (see Chapters "Diet composition and trophic ecology of Northeast Pacific Ocean sharks" by Bizzarro et al., this volume; "Stable isotope applications for understanding shark ecology in the Northeast Pacific Ocean" by Reum et al., this volume). Historically, sharks were directly targeted for oil primarily from their large livers and were also systematically removed from nearshore marine waters because they were considered a nuisance to fishers when they became entangled in nets set for other species or stripped bait from recreational hook-and-line gear. Owing to their life history characteristics, many species of shark cannot sustain extended periods of such intensive and direct take without negatively impacting the health of their populations and evolutionary potential (Dulvy et al., 2014).

A clear example of overexploitation and slow recovery following excessive take of a NEP shark is the Basking Shark, *Cetorhinus maximus*. Schools of hundreds to thousands of these sharks used to roam the nearshore waters of the NEP off of California, Washington, and British Columbia (McFarlane et al., 2009; Squire, 1990). From the 1950s to the 1970s, directed Basking Shark fisheries for oil and shark eradication programs, originally implemented to reduce impacts upon commercial salmon fishing gear, depleted the population to an estimated 10% of preexploitation levels (McFarlane et al., 2009; Squire, 1990). Today, Basking Shark sightings are rare in the NEP despite decades of low fishing pressure; their recovery is likely hindered by their slow population growth rates as well as continued take from illegal shark finning, bycatch from other fisheries, and boat strikes in shallow coastal waters. Levels of mitochondrial DNA (mtDNA) diversity in Basking Sharks worldwide is low (nucleotide diversity = 0.001) with no genetic difference in populations between ocean basins, potentially limiting the evolutionary potential of this species (Hoelzel et al., 2006).

In this review, we explore the approaches, results of studies to date, and future applications of molecular tools in shark research and discuss the implications that conservation genetics have for the management of NEP sharks.

2. APPROACHES

2.1 Tissue Sampling

Sample types used in conservation genetic studies of sharks to date include blood and small skin or muscle biopsies from captive and live-caught animals, as well as similar tissues from stranded dead animals. Fresh tissue samples preserved in 70%–100% ethanol, any other commercially available DNA preservative, or frozen immediately after sampling from live or freshly dead animals provide the greatest quantity of intact DNA. Blood samples from captive or live-caught sharks may yield high-quality DNA extracted from nucleated red blood cells. Recently, environmental DNA (eDNA) isolated from water samples has been used to determine presence/absence of target fish species in a given environment using assays that can identify trace amounts of DNA shed from live animals (Kelly et al., 2014). Skin or muscle samples that are not freshly collected and preserved may still yield DNA, but it tends to be lower in quality and quantity. For example, historic samples such as museum specimen (dry or formalin preserved), trophies (e.g. shark jaws, sawfish rostra), and teeth can also yield DNA providing valuable information about past population structure and diversity (Gubili et al., 2015). Since the DNA in such samples tend to be of low quality and quantify, studies incorporating historic samples typically utilize shorter reads of mtDNA or nuclear microsatellites (highly variable repeating nucleotide regions) rather than next-generation sequencing techniques looking at thousands of sites genome-wide (Etter et al., 2011). To date, skin and muscle tissues are by far the most commonly used sample types in NEP shark genetics studies (see studies in Table 1).

Quantifying the variation in proteins from a wide range of tissues has been used in the last 2 decades of the 20th century to characterize taxonomic relationships, population structure, and gene function in sharks, longer than any other genetic tool. Molecules such as allozymes (enzymes or proteins used as molecular markers) are relatively uninformative compared with present-day nucleotide sequencing, and more expensive to develop, and are therefore slowly giving way to more current tools like whole-genome sequencing that can yield hundreds to thousands of data points per sample. Yet allozymes and other forms of protein electrophoresis still have modern applications, especially in identifying individual animals when microsatellites are unavailable, and in characterizing the expression of coding gene regions (Li et al., 2013; Venkatesh et al., 2014). The sequencing of the Elephant Fish, *Callorhynchus millii*, genome has especially been a boon to such study of chondrichthyes (Venkatesh et al., 2014), providing the first full reference library for a cartilaginous fish and a potent tool for comparing gene function across divergent taxa (Chen et al., 2012; Opazo et al., 2015; Venkatesh et al., 2014).

2.2 Mitochondrial DNA

Mitochondrial DNA (mtDNA) is a nonrecombinant circular molecule of nucleotides whose genes primarily code for proteins used for gene transcription and cellular respiration. In sharks, as in mammals, mtDNA is inherited solely through matrilines, and its rate of molecular evolution, which ranges from relatively slow in the cytochrome oxidase I gene, or barcoding gene, to relatively fast in the D-loop segment of the mitochondrial control region (CR), have made mtDNA a useful tool for phylogenetics and phylogeography for a number of years (Avise, 1994). Though much focus has recently switched to rapid whole-genome sequencing, studies that utilize mtDNA continue to be relatively inexpensive to develop and highly informative, and the number of published whole shark mitochondrial sequences continues to grow. Though cytochrome oxidase I (CO1) and the CR have found wide use in studying both species- and population-level

Table 1 Northeast Pacific Shark Genetic Studies, Methodology, and Results

Species	Region Sampled	Number of Individuals	DNA Marker Used	Mitochondrial Diversity (±SD)	Nuclear Diversity (±SD)	References
Alopias pelagicus (Pelagic Thresher Shark)	Northern and Southern EP	121	mtDNA (CO1)	$h = 0.5866 \pm 0.036$; $\pi = 0.104$	NM	1. Cardeñosa et al. (2014)
	Southern California	11–19	Microsatellites (9 loci)	NM	Ho = 0.875–0.944 (avg. = 0.919)	
Carcharodon carcharias (White Shark)	Northern EP	59	mtDNA (CR)	$\pi = 0.00131 \pm 0.00089$	NM	2. Jorgensen et al. (2009)
Galeorhinus galeus (Tope Shark)	Northern EP	26	mtDNA (CR)	$h = 0.73 \pm 0.08$; $\pi = 0.0019 \pm 0.0012$	NM	3. Chabot and Allen (2009)
Hexanchus griseus (Sixgill Shark)	Northern EP (Washington State)	51	mtDNA (CO1, CR, and ND2)	$h = 0.000$; $\pi = 0.000$	NM	4. Daly-Engel et al. (2017)
		>300	Microsatellites (10 loci)	NM	$A_r = 8.00$	5. Larson et al. (2011)
Isurus oxyrinchus (Shortfin Mako)	Northern EP	30	mtDNA (RFLP)	$h = 0.861$; $\pi = 0.381$	NM	6. Heist et al. (1996)
	Worldwide (including NEP)	421–430	Microsatellites (4 loci)	NM	Ho = 0.77–0.91 (avg. = 0.848)	7. Schrey and Heist (2003)
	Northern EP	90	Microsatellites (5 loci)	NM	Ho = 0.78–0.96 (avg. = 0.89)	8. Corrigan et al. (2015)

Continued

Table 1 Northeast Pacific Shark Genetic Studies, Methodology, and Results—cont'd

Species	Region Sampled	Number of Individuals	DNA Marker Used	Mitochondrial Diversity (±SD)	Nuclear Diversity (±SD)	References
Mustelus henlei (Brown Smoothhound Shark)	Northern EP	126	mtDNA (CR)	$h = 0.77 \pm 0.04$; $\pi = 0.004 \pm 0.003$	Ho = 0.45; $A_r = 4.11$	9. Chabot et al. (2015)
		169				
Notorynchus cepedianus (Sevengill Shark)	Northern EP	130	Microsatellites (7 loci)	NM	Ho = 0.171–0.593 (avg. = 0.417)	10. Larson et al. (2015)
Rhincodon typus (Whale Shark)	Northern EP	8	mtDNA (CR)	$h = 0.96 \pm 0.08$; $\pi = 0.006 \pm 0.004$	NM	11. Castro et al. (2007)
	Tropical EP	10	Microsatellites (8 loci)	NM	$A_r = 2.000–7.878$ (avg. = 4.181)	12. Schmidt et al. (2009)
Sphyrna lewini (Scalloped Hammerhead Shark)	Baja	44	mtDNA (CR)	$h = 0.51 \pm 0.018$; $\pi = 0.0009 \pm 0.0009$	NM	13. Duncan et al. (2006)
	Northern EP	61	Microsatellites (13 loci)	NM	$A_r = 6.91$	14. Daly-Engel et al. (2012)

Squalus suckleyi (Spiny Dogfish)	Northern EP (Washington, Oregon, and California)	94	mtDNA (ND2)	h=0.89–0.93 (avg.=0.91); π=0.0048–0.0054 (avg.=0.0051)	NM	15. Verissimo et al. (2010)
		99	Microsatellites (8 loci)	NM	Ho=0.545	
Triakis semifasciata (Leopard Shark)	Northern EP (California)	169	mtDNA (CR); intersimple sequence repeats (ISSRs)	π=0.0067	He=0.109±0.003	16. Lewallen et al. (2007)
		377	Microsatellites (5 loci)	NM	A_r=4.13–5.57 (avg.=5.03)	17. Barker et al. (2015)

CO1=cytochrome oxidase 1; CR=control region; ND2=NADH-dehydrogenase 2; NM=not measured; h=haplotype diversity; π=nucleotide diversity; A_r=allelic richness; Ho=observed heterozygosity; He=expected heterozygosity (not reported unless observed heterozygosity was unavailable).

genetic variation in sharks, other mitochondrial genes with intermediate mutation rates, such as the NADH dehydrogenase-2 and -4 (ND2 and ND4) genes, are growing in popularity for these purposes. Analysis of such loci can also shed light on contemporary sex-differentiated movement patterns and shark reproductive behaviour, when data from mtDNA markers are compared with those from biparentally inherited nuclear genes or telemetry data to elucidate sex-biased migration patterns and differentiate physical from genetic dispersal (Chabot, 2015; Chabot and Allen, 2009; Chabot et al., 2015; Daly-Engel et al., 2012; Jorgensen et al., 2009; Pardini et al., 2001; Prugnolle and de Meeus, 2002).

DNA barcoding, a method of species identification through the sequencing of short fragments of mtDNA, has garnered particular attention in recent years (Hebert et al., 2003; Taylor and Harris, 2012). The locus typically used in DNA barcoding in sharks is CO1, which is instrumental for the regulation of cellular respiration and therefore has a relatively slow rate of molecular evolution (Hebert et al., 2003). This locus has little within-species diversity but is useful for identifying divergence between species, with primer sequences that are often conserved across shark genera and higher taxonomy, making the use of universal primers possible (Krishnamurthy and Francis, 2012; Taylor and Harris, 2012). Since the mid-2000s, DNA barcoding has been used heavily in the identification of invertebrates, plants, fish, bird, and other species (Taylor and Harris, 2012). As a result of concerted efforts, databases like GenBank (http://www.ncbi.nlm.nih.gov/genbank/), the Barcode of Life Data Systems (http://www.boldsystems.org/), and The Fish Barcode of Life Initiative (FISH-BOL; http://www.fishbol.org/) now contain valuable reference libraries of CO1 and other genes from a wide variety of organisms, including many elasmobranchs. In sharks, DNA barcoding has been used to identify cryptic species (Ovenden et al., 2009; Ward et al., 2005), species caught in commercial fisheries and sold in fish markets (O'Bryhim et al., 2017), and species or population origin of fins from illegal, unreported, and unregulated (IUU) fishing vessels (Ward et al., 2008). There are concerns, however, over the misidentification of hybrids, since DNA barcoding utilizes only mtDNA, and therefore would only identify the maternal species in sharks (Ward et al., 2008).

Recently, studies utilizing mtDNA data in sharks of the NEP have brought significant insights into shark coastal ecology and evolution, particularly among colder-water species that are often inaccessibly deep ranging in other parts of the world, such as Sixgill Sharks, *Hexanchus* spp., or highly migratory coastal-pelagic species that make ample use of the NEP's productive waters, such as Mako Sharks, *Isurus* spp. and White Sharks, *Carcharodon carcharias* (Heist et al., 1996; Jorgensen et al., 2009; Pardini et al., 2001). The ubiquity of the CO1 gene and others for DNA barcoding, coupled with the ease of Sanger sequencing relative to the processing and bioinformatics required for most whole-genome sequencing, means that mtDNA analysis is likely to persist as a useful tool in elasmobranch genetics for some time.

2.3 Microsatellites

Microsatellites are nuclear markers that gained popularity during the 1990s because they are highly variable relative to mtDNA and abundant within the nuclear genome (Bentzen et al., 1991, Park and Moran, 1994, Wright and Bentzen, 1995). Microsatellites are stretches of DNA with repeating units of 1–6 base pairs (bp), assumed to be functionally neutral (but known to act in some human diseases; Chen et al., 2002, Paulson et al., 1996), and are easily amplified and assayed in any conventional molecular laboratory using polymerase chain reaction (PCR) (Li et al., 2004; Thibodeau et al., 1993). Microsatellites are codominantly inherited, with both parents contributing alleles to the next generation, and can mutate rapidly via slip-strand mispairing in germ cells during DNA replication (Avise, 1994). Such high polymorphism across multiple loci can be used as a 'genetic fingerprint' to identify individuals, making microsatellites ideal for studying kinship, relatedness, genetic mating systems, and identifying relatively recent population structure (Selkoe and Toonen, 2006). Some limitations to the applicability of microsatellites include the following: (1) they are most effective when species-specific, and although they are often conserved within genera and even families, they may lose variability and thus effectiveness; (2) the mutation model remains unclear (infinite alleles vs stepwise mutation), complicating the statistical assumptions used to predict allelic variability; (3) they can carry null alleles (alleles that are not amplified because of a point mutation), causing difficulties in statistical models (Selkoe and Toonen, 2006); and (4) they are expensive and somewhat time consuming to develop and optimize for new species. Nevertheless, microsatellites and mtDNA sequence variation are currently the most widely used molecular markers in shark genetics studies (Table 1).

2.4 MHC Genes

Genes of the major histocompatibility complex (MHC) have been widely used in population genetics to determine immune diversity, phylogenetics, and adaptive potential (Garcia de Leaniz et al., 2007). The MHC genes are categorized into three groups: Classes I, II, and III (Sugaya et al., 1994).

Expression of MHC genes is influenced by environmental change and exposure to invaders or disease vectors, making them particularly useful in highly variable coastal marine environments. In sharks, MHC genes have been used to indicate directional selection from a variety of sources, including sexual selection and other factors that can impact the recombination of genomes (Bartl and Nonaka, 2014; Edwards and Hedrick, 1998). MHC genes have also been useful in deep phylogenetics, indicating coevolution among some lineages of basal vertebrates (Bartl, 1998; Kulski et al., 2002; Ohta and Flajnik, 2015; Okamura et al., 1997), and likely monophyly among sharks in at least some MHC classes (Shen et al., 2014). MHC genes have also shed light on elasmobranch reproductive strategies, as brood sampling to examine MHC evolution led to the first findings of multiple paternity in a shark species (Fitzpatrick et al., 2012; Ohta et al., 2000; Saville et al., 2002).

2.5 Single-Nucleotide Polymorphisms (SNPs)

For the past 25 years, microsatellite loci and mtDNA sequence data have been the predominant markers in population genetics research. However, with recent advances in technology and decreasing costs of generating genome-wide datasets using next-generation sequence techniques, SNPs have become increasingly popular. SNPs are single base-pair substitutions distributed throughout the nuclear and mtDNA genomes in both coding and noncoding regions (Morin et al., 2004). SNPs are often scored more easily than microsatellite loci because they lack shift mutations (Morin et al., 2004) and are the traditional locus of choice scored in single-gene sequencing. However, in contrast to microsatellite markers, SNPs tend to be biallelic and the rate of mutation at each site is low (Jones et al., 2009; Morin et al., 2004), requiring a large number of loci and long stretches of genome for population-level analyses: an estimated 4–10 times more SNPs are likely to be necessary to assess levels of genetic diversity compared with microsatellite markers, and 40–100 times may be required to assess relatedness (Jones et al., 2009; Mariette et al., 2002; Morin et al., 2004). Analysing hundreds to thousands of SNPs generated with next-generation sequencing methods requires relatively high processing power and knowledge of bioinformatics, resources not yet available to all researchers. Thus SNPs analysed without the traditional Sanger sequencing methods used for single-gene mtDNA studies are still largely untried in conservation genetic studies of NEP sharks. This technology is rapidly becoming more inexpensive and accessible, however, and has been successfully applied in sharks elsewhere, and other marine fishes (Krück et al., 2013; Milano et al., 2014; Portnoy et al., 2015). New methods for analysis are being developed at a high rate of speed, and protocols developed within the last few years, especially for the highly useful technique of restriction site-associated DNA sequencing (RAD-seq; reviewed in Blair et al., 2015). It is likely that in the next decade, whole-genome SNPs will become a more commonly used marker type in conservation genetic studies on elasmobranchs, possibly replacing microsatellite loci in effectiveness, and popularity for assessments of adaptation and recent population structure.

2.6 Forensic Genetics

Wildlife forensics is an area of conservation genetics that focuses on the use of molecular tools to provide DNA evidence for monitoring the wildlife trade and enforcing management laws (Ogden et al., 2009). Research in wildlife forensics is often aimed at developing and implementing forensics tools to identify a species, population, or the individual identity of biological material (Ogden et al., 2009). International trade in several shark species is banned under Appendix I and regulated under Appendix II of the Convention on International Trade in Endangered Species (CITES), and some species of shark are protected within US waters. However, monitoring the trade of products and/or parts from these protected species has long been obstructed by difficulties in species identification because the whole animal is not available (Magnussen et al., 2007), so the potential value of genetic identification from small or trace tissue samples is high. Approaches like DNA barcoding have been effectively used to identify species or even populations of sharks caught in illegal fisheries and sold in fish markets, both in the United States and overseas, in an effort to enforce protective laws and to monitor the catch and trade of shark parts (Hoelzel, 2001; Magnussen et al., 2007; O'Bryhim et al., 2017).

3. STUDIES TO DATE

3.1 Proteins

Recent protein analysis has shed light on physiology and speciation in elasmobranchs and other fishes (Buonocore and Gerdol, 2016; Goshima et al., 2016), including the existence of an advanced complement system that enhances innate immune function in sharks (Goshima et al., 2016) and the role of enzymes in shark digestion (Jhaveri et al., 2015). These and other findings, such as detection of cyanobacterial neurotoxins in shark tissues (Mondo et al., 2012), adaptation to temperature variation and the origins of endothermy (Ciezarek et al., 2016), and the rapid identification of stress indicators such as corticosteroids and blood glucose (Hoffmayer et al., 2015; Skomal and Mandelman, 2012) even after death (Wosnick et al., 2017), are particularly applicable in the face of increased attention to the effect of climate change on marine populations. In the NEP, investigation within the past decade has identified the role that specific proteins may play in adaptation to salinity change in Leopard Sharks, *Triakis semifasciata* (Dowd et al., 2010), and North Pacific Spiny Dogfish, *Squalus suckleyi* (Deck et al., 2016; Guffey and Goss, 2014). Antioxidant enzymes found in Blue Sharks, *Prionace glauca*, in Baja California have shed light on habitat and trophic variation among conspecific cohorts (Barrera-García et al., 2012), and the identification of the role that heat shock proteins play in thermal adaptation (Marshall et al., 2012) among the Common Thresher Shark, *Alopias vulpinus*, populations of California (Heberer et al., 2010) are likely to prove instrumental in advancing our understanding of how sharks may react to climate change. The impact of increasing sea surface temperatures, ocean acidification, prey shifts, and other factors associated with climate change, have largely unknown effects on sharks in the NEP. The NEP is a complex oceanographic region with some areas more vulnerable and some areas more resilient to climate change because of multiple and complex ocean currents, variable benthos (both shallow and deep areas), the presence of large estuaries (San Francisco Bay, Willapa Bay, The Salish Sea, Glacier Bay, and Prince William Sound), and many other complicating factors. In addition, the impacts of climate change likely differ depending on each shark species' life history, habitat preferences, and ecology (O'Brien et al., 2013).

3.2 Genetic Diversity

Genetic variation among individuals, populations, and species is thought to be critical for population adaptability and survival in response to changing environmental conditions, such as climate change and degraded habitats. Genetic diversity is a critical measure in population genetics because it can inform on the genetic health of a population. For example, low levels of genetic diversity over the long term may reduce population fitness and evolutionary potential because individuals may not have the genetic flexibility to respond to changing climate conditions or disease.

Genetic diversity in sharks may naturally be lower than within other vertebrates. Shark genomes change or evolve much slower than in other organisms such as mammals and bony fishes, likely due to their slower metabolic rates (Martin, 1999). In comparison to mammals, shark mtDNA substitution rates are seven to eight times slower and nuclear genes substitution rates are 10–20 times slower (Martin, 1999; Martin et al., 1992). Because of this relatively slow rate of change in shark genomes, shark populations may take a long period of time to recover from historic (evolutionary) bottlenecks and to genetically diverge, even in the absence of gene flow. This slow rate of mutation also makes it difficult to differentiate between recently diverged species, as the accumulation of genetic differences between species is an evolutionarily slow process (Martin et al., 1992).

Genetic diversity has been described in several species of sharks in the NEP using data from mtDNA markers and microsatellite loci. Genetic diversity using mtDNA markers is described using haplotype and nucleotide diversities. Haplotype (allelic) diversity (h) represents the probability that two randomly selected individuals exhibit different haplotypes while nucleotide diversity (π) represents the probability that two randomly selected homologous nucleotides are different (Nei, 1987). Genetic diversity at microsatellite loci is typically described using the proportion of polymorphic loci, heterozygosity, and allelic diversity (Frankham, 2010). Heterozygotes are defined as individuals that carry two different alleles or versions of the gene being studied, one from the sire and one from the dam, while homozygotes carry two copies of the same allele. Heterozygosity is calculated as a proportion of heterozygotes in the sampled population, and can be expressed as both observed heterozygosity (Ho) and expected heterozygosity (He) based on allelic variation expected under Hardy–Weinberg equilibrium—a model that assumes allele frequencies in a population will remain constant from generation to generation (Crow, 1999). Allelic diversity or richness is the measurement of the mean number of alleles per locus (genetic location or gene) across the sampled population.

Assessments of genetic diversity in NEP sharks to date have utilized mtDNA markers and/or highly polymorphic species-specific microsatellites isolated from their nuclear genomes (Table 1). Levels of nucleotide diversity in NEP sharks are typically low, a trend observed in many sharks, likely due to the slow rate of mtDNA mutation in this taxonomic group (Table 1; Martin et al., 1992). Levels of haplotype diversity in NEP sharks range from very low to high (Table 1). For example, the NEP Sixgill Shark (Daly-Engel et al., 2017) and Whale Shark, *Rhincodon typus* (Schmidt et al., 2009), have extremely low levels of haplotype diversity, although the sample size was very small for the latter (Table 1). The Pelagic Thresher Shark, *Alopias pelagicus*, (Cardeñosa et al., 2014) and Scalloped Hammerhead Shark, *Sphyrna lewini*, (Duncan et al., 2006) have moderate levels of haplotype diversity while the Shortfin Mako, *Isurus oxyrinchus* (Heist et al., 1996), Spiny Dogfish (Verissimo et al., 2010), Tope Shark, *Galeorhinus galeus* (Chabot and Allen, 2009), and Brown Smoothhound Shark, *Mustelus henlei* (Chabot et al., 2015), have relatively high levels of haplotype diversity (Table 1).

NEP shark species with relatively high genetic diversity at microsatellite loci include the Scalloped Hammerhead, the Shortfin Mako, and the Leopard Shark with average Ho ranging from 0.873 to 0.703 and He ranging from 0.873 to 0.754 (Barker et al., 2015; Daly-Engel et al., 2012; Schrey and Heist, 2003). Those with moderate genetic diversity at microsatellite loci are the Pelagic Thresher Shark, Whale Shark, and Bluntnose Sixgill Shark, *Hexanchus griseus*, with average Ho ranging from 0.660 to 0.450 and He ranging from 0.694 to 0.610 (Cardeñosa et al., 2014; Larson et al., 2011; Schmidt et al., 2009). Those with lower genetic diversity at microsatellite loci are the Nurse Shark, *Ginglymostoma cirratum*, the Blue Shark, the North Pacific Spiny Dogfish, and the Broadnose Sevengill Shark (*Notorynchus cepedianus*), with average Ho ranging from 0.600 to 0.550, and He 0.560 to 0.540 (Larson et al., 2015; Ovenden et al., 2009; Verissimo et al., 2010), the Tope Shark with average Ho = 0.34 and He = 0.40 (Chabot, 2015), and the Brown Smoothhound Shark with average Ho = 0.45 and He = 0.56 (Chabot et al., 2015). The lower levels of diversity found in the latter benthopelagic sharks may be due to historical population reductions and/or contemporary overexploitation in these regions, though may also be indicative of the lower metabolic and mutation rates typically observed in the deeper-water sharks (O'Brien et al., 2013).

3.3 Population Structure and Dispersal

Assessments of population structure depict the connectivity and/or genetic distinctness of groups of individuals in different locations and may, therefore, inform on the appropriate spatial scale for management decisions. While the goal in population structure studies is often to define population boundaries, in practice this is difficult due to the nature of such boundaries and the feasibility of sampling (Lewallen et al., 2007). Whether or not the boundaries between populations are relatively discrete, resemble an isolation-by-distance pattern, or another pattern entirely often remains unclear (Wright, 1943). Over the past decade, assessments of population structure in elasmobranchs have primarily employed mtDNA and microsatellite markers. Most shark species studied to date exhibit matrilineal structuring, although the spatial scale of the structuring ranges from between ocean basins in pelagic species to along continuous coastlines in more coastal or benthic species (Chabot and Allen, 2009; Chabot et al., 2015; Daly-Engel et al., 2012; Dudgeon et al., 2009; Karl et al., 2012; Keeney et al., 2005; Phillips et al., 2017). The presence of matrilineal structuring is usually attributed to the philopatric behaviour of females to parturition sites that are suitable for young, increasing their chances of survival (Chapman et al., 2015; Feldheim et al., 2014). In some species, such as the Lemon Shark, *Negaprion brevirostris*, this philopatry is highly localized, with females homing to their natal nursery for parturition (Feldheim et al., 2014). It has proven more difficult to find patterns in the gene flow of male elasmobranchs, requiring a comparison between matrilineal and biparental markers, or sex data recorded at the time of capture. Whether males disperse widely or maintain a home range within which mating takes place tends to vary among species (Phillips et al., 2017).

Only a handful of population structure studies have included sample sites in the NEP, and very few have specifically focused on resolving small-scale population structure in this region (Table 1; but see Chabot et al., 2015). In many large, highly vagile pelagic and coastal-pelagic species, the matrilineal lineages are genetically differentiated between oceans, across ocean basins, and between northern and southern regions of an ocean. For example, White Sharks have mtDNA structuring between the north Pacific (i.e., California) and both the southwest Pacific (i.e., New Zealand) and South Africa (Jorgensen et al., 2009). Patterns of gene flow among male sharks often differ relative to that of the females, with male gene flow occurring across ocean basins and sometimes circumglobally. The matrilineal lineages of Shortfin Mako in the NEP are genetically distinct from those in the north and south Atlantic, but there appears to be no barrier to male gene flow between or within these ocean basins (Heist et al., 1996; Schrey and Heist, 2003). The Whale Shark lacks mtDNA structuring between the Pacific and Indian oceans, but this population is distinct from those in the Atlantic Ocean, while males appear to disperse readily between the Pacific, Indian, and Atlantic oceans (Schmidt et al., 2009). Female Scalloped Hammerheads tend to be philopatric and closely linked to coastal pupping sites, whereas male gene flow occurs across ocean basins (Daly-Engel et al., 2012;

Duncan et al., 2006). Contrary to this generalized pattern of male gene flow in Scalloped Hammerheads, however, a single site in the NEP was genetically distinct from all other sites in the Pacific (Hawaii and Western Pacific), perhaps due to vast open water barrier separating the NEP from other biogeographic regions (Daly-Engel et al., 2012). The Tope Shark is genetically structured between the eastern Atlantic Ocean, eastern Pacific Ocean and Australia, and also at smaller spatial scales within ocean basins; between the NEP and southeastern Pacific and Atlantic oceans at both mitochondrial and nuclear loci (Chabot, 2015; Chabot and Allen, 2009).

Species that are smaller, less vagile, and benthic or benthopelagic tend to exhibit population structure at smaller spatial scales. For example, the Brown Smoothhound shark in northern California, central-southern California/Mexico, and Costa Rica are genetically distinct (Chabot et al., 2015). Similarly, the Leopard Shark is genetically structured in the NEP over small spatial scales relative to larger pelagic species like those mentioned above (Barker et al., 2015; Lewallen et al., 2007). Populations of Leopard Shark in northern California and southern California are genetically differentiated, with a pattern of isolation-by-distance (Barker et al., 2015; Lewallen et al., 2007). In addition, Humboldt Bay appears to contain a genetically and ecologically unique population, although there may be some gene flow with individuals from San Francisco Bay to the south (Barker et al., 2015; Lewallen et al., 2007). In the Nurse Shark, populations are subdivided within the Atlantic Ocean basin, and although it was not formally tested due to the small sample sizes obtained in the Pacific Ocean, it is probable that Nurse Sharks in the NEP are genetically differentiated from those in the Atlantic and possibly within the Pacific itself (Karl et al., 2012). In at least one instance, population differentiation has led to phylogenetic redescription: the Spiny Dogfish Shark once assumed to be a single circumglobal species, was found to constitute a distinct species (*S. suckleyi*) in the NEP portion of its range (Ebert et al., 2010; Hauser, 2009; Verissimo et al., 2010).

3.4 Reproductive Strategies

Reproduction can have a profound impact on rebound potential and population viability, and fecundity can have a substantial impact on conservation in large, slow-growing organisms such as sharks. As mentioned previously, strategies such as male-biased dispersal and female philopatry to coastal nursery habitat are ubiquitous in most temperate shark species studied to date, including those whose range includes the NEP. Therefore, molecular investigation of shark and other elasmobranch species using primarily mtDNA and microsatellites has expanded our understanding of reproduction-based movement patterns and mating strategies, with corresponding implications for fisheries management. In recent years, the NEP has been found to comprise critical reproductive habitat that supports populations of neonate, juvenile, and subadult sharks belonging to a number of important pelagic and coastal-pelagic fisheries species. In Common Thresher Sharks, *A. vulpinus*, juveniles inhabit continental shelves off of California and Mexico through much of the year (Cartamil et al., 2016), potentially rendering this stock vulnerable to depletion through coastal overfishing from both the United States and Mexico. In addition, the Salish Sea has been found to be a year-round nursery area for subadult Bluntnose Sixgill Sharks (Larson et al., 2011). Genetic tools have been used to document the extensive use of the NEP by White Sharks for regular seasonal aggregations that are presumably linked to their 18-month gestation period (Jorgensen et al., 2009), resulting in the identification of mating and pupping areas that connect populations in the NEP with locations in the south and west Pacific (Domeier and Nasby-Lucas, 2013).

Molecular tools have also been used widely and successfully to elucidate shark genetic mating systems, particularly the practice of polyandry (females mating with multiple males over the course of a single breeding season). Though promiscuous mating by male sharks (polygyny) has been observed directly (Pratt and Carrier, 2001; Whitney et al., 2004), polyandry has only been brought to light by revealing multiple paternity (when a brood of offspring from a single dam is sired by multiple males) using RFLP or microsatellites, either through direct brood sampling or assignment-based testing and analysis of kinship and relatedness (Byrne and Avise, 2012; Feldheim et al., 2004; Fitzpatrick et al., 2012; Larson et al., 2011). In the NEP, this work has taken the form of opportunistic sampling of a few broods from large-bodied sharks, and through more thorough sampling of multiple broods from smaller shallow-water species. As previous reviews on the topic have shown, high variation in frequency of multiple paternity between species and populations appears to be the norm for sharks, though the source of such variation is poorly understood (Daly-Engel et al., 2010; Fitzpatrick et al., 2012). Multiple paternity was found in single litters of Bluntnose Sixgill Sharks (Larson et al., 2011) and Shortfin Mako (Corrigan et al., 2015) in the NEP, indicating that this strategy is potentially common in these species, and may be the mechanism responsible for the high relatedness observed in NEP populations of sixgills and Broadnose Sevengill Sharks due to a higher number of half-siblings among young of the year than in monogamously mating populations (Larson et al., 2011, 2015). In smaller species, more exhaustive sampling is possible, leading to estimation of frequency of multiple mating as opposed to occurrence. Frequency of multiple paternity averaged 36.4% in California Leopard Sharks, though wide temporal variation was observed over the 2 years of the study (20.0%–83.3%) (Nosal et al., 2013), possibly due to the proportion of aggregate males present at the time of mating. Similar variation was observed in the Brown Smoothhound Shark though in this case frequency of multiple paternity ranged from 0% to 93.3% and varied between populations as well as sampling years (Chabot and Haggin, 2014).

Though genetic tools have been instrumental in identifying reproductive behaviour and migration, it continues to be important for management purposes to link these data with information gleaned from direct observation and telemetry, in order to best document movement patterns of animals that may occur with or without the presence of gene flow. Furthermore, while the use of next-generation sequencing to screen entire genomes of SNPs has found wide application in understanding adaptive evolution and broad-scale allele frequency shifts, microsatellites continue to be the most effective and easily applied tool in the genetic arsenal for characterizing mating behaviour like polyandry and relatedness, at least for now. Other reproductive strategies that can be elucidated through the use of genetic tools, such as parthenogenesis or 'virgin birth', have been documented in sharks (e.g., Chapman et al., 2008; Robinson et al., 2011) but not yet in the NEP.

4. FUTURE WORK

4.1 Genetics of Adaptive Fitness

Genetic studies not only infer fitness through the measurement of genetic diversity, but can also provide valuable insights into genetic adaptation; how sharks may adapt to environmental challenges such as climate change, prey shifts, and exposure to novel pathogens. One type of adaptive genetics examines sections of the genome associated with immune function and how the shark's immune system responds to threats to their health. Acquired immunity is engaged if the invading agent is not inactivated or killed by innate mechanisms, can be specific to the invading agent, and includes long-term cellular 'memory'. Specificity is enabled through recognition of particular proteins or antigens, and memory is developed so repeated exposure to the same invading agent will elicit a rapid response of antigen-specific cells to destroy the invader. The immune system of sharks is similar to that of other vertebrates including warm-blooded animals such as humans, which utilizes cell-mediated immune responses together with the MHC genes (Criscitiello et al., 2006). To date, no NEP sharks have been examined for MHC gene variation.

The genetics of adaptation can also be measured via gene expression: the conversion of the information encoded in a gene into messenger RNA (mRNA) and then into a protein. Variability is calculated by measuring the mRNA, and thus indirectly the protein to be expressed. Gene expression is influenced by environmental changes, exposure to disease, contaminants, and other external variables such as prey availability and reproductive status. New technology has enabled mass sequencing of transcriptomes similar to whole-genome sequencing, but there too costs remain high and utility in sharks is limited as yet. One study on the White Shark from the western Atlantic looked at gene expression and discovered the heart transcriptome in White Sharks to be more similar to humans than to Zebrafish, *Danio rerio* (Richards et al., 2013). To date, there have been no studies of gene expression in sharks of the NEP.

4.2 Trace DNA

Noninvasive genetic sampling is highly valuable for conservation genetic studies where minimizing injury and stress to a threatened or endangered species is a high priority. Active (as opposed to opportunistic) genetic sampling protocols for sharks typically involve setting nets or lines, capturing the target species, and handling the animal to remove a small piece of tissue. For larger species, such as the Whale Shark, genetic samples are usually obtained via hand spears armed with biopsy darts while diving (Castro et al., 2007). Such methods are not only costly and time consuming, but inflict varying amounts of trauma to the animal, with potentially harmful results. Besides impacting both target and nontarget species, sharks have been shown to be prone to stress-related morbidity, with postrelease recovery and survival varying widely among species (Frick et al., 2010; Gallagher et al., 2014). In addition, capturing, handling, and/or directly interacting with protected species require special permits that can be time consuming or challenging to obtain. Fortunately, DNA technology is becoming increasingly more sensitive, allowing for the use of less biological starting material. Two of the most promising methods of noninvasive sampling are skin swabbing and eDNA, sometimes called 'trace' DNA. Mucus and skin cells can be collected by rubbing the surface of the skin lightly, and this has been demonstrated to successfully yield DNA of sufficient quality and quantity to generate data at both mtDNA and microsatellite markers (Kashiwagi et al., 2015; Lieber et al., 2013). The use of eDNA takes noninvasive sampling one step further, negating the need to capture or even observe the target species. All aquatic species, including sharks, leave traces of their DNA in the environment via shed cellular debris, skin cells, metabolic waste, and blood, all of which can be collected in water and/or sediment samples. DNA is extracted from these environmental samples, and species-specific primers are developed to amplify small DNA fragments in only the target species. Initial eDNA studies on sharks, similar to ones done on bony fishes, have focused on providing an alternative method of conducting species surveys with simple presence/absence data, such as that used in the study of Largetooth Sawfish, *Pristis pristis* (Simpfendorfer et al., 2016). In the near future, eDNA methods will be used to study shark abundance, biodiversity, and answer population-level questions (Pilliod et al., 2014; Port et al., 2016; Sigsgaard et al., 2016).

While eDNA holds much promise, using appropriate methodologies is essential for valid results. For example, when using eDNA to survey for the presence of a particular species in an area, primer specificity is key (Wilcox et al., 2013). Primers should not only be tested for positive amplification in the target species, but they should also be tested in closely related, sympatric species to eliminate the potential for cross-species amplification and false positives (Wilcox et al., 2013). Negative results also require careful consideration; a negative could indicate the species is not present, but could also result from target DNA concentrations in samples below the detection thresholds of the methods employed. In order to improve the accuracy of the results of eDNA studies, methods that are sensitive to minute amounts of DNA and employ quantitative approaches like quantitative PCR (qPCR), which can estimate the quantity of template molecules in the original sample. In addition, baseline research is needed specifically assessing the residency time of elasmobranch DNA in different systems (e.g., open ocean, coastal, and estuarine) to better understand the relationship between environmental conditions and eDNA persistence.

4.3 Epigenetics

Organisms and their traits change and adapt in response to their environment. Epigenetics research investigates changes in organisms which are caused by modifications in gene expression, as opposed to changes in DNA sequences. These modifications may include the methylation of cytosine in the DNA, chemical modification of chromatin structure, or regulatory processes mediated by RNA (Bossdorf et al., 2008). Epigenetics has shown great promise in documenting rapid changes in adaptation via gene expression in other organisms, including fishes (Veilleux et al., 2015), however very little work has been done to date on elasmobranchs. The only known study to date, on the Winter Skate, *Leucoraja ocellata*, from the Canadian Atlantic, showed that population expansion within the last 7000 years has resulted in transcriptome-wide epigenetic changes without a corresponding nucleotide sequence change (Lighten et al., 2016). These changes likely occurred in response to the 10 °C decrease in water temperature that accompanied a habitat shift in one portion of the skate's range, resulting in differential expression of cold tolerance genes (Lighten et al., 2016). Studies such as this, demonstrate the potential that epigenetics have to help pinpoint the mechanisms of climate adaptation, especially in slow-growing shark species of the NEP that experience low baseline levels of genetic diversity but high potential for environmental change.

5. CONCLUSION

The conservation genetics of NEP sharks remain understudied, with great potential for future research. The genetic tools described here have thus far been used to assess population structure, demography, levels of genetic diversity and reproduction in NEP sharks, which would have been very difficult to investigate using other methods. Newer genetic methods such as whole-genome sequencing, adaptive genetics, and epigenetics, together with the use of noninvasive sampling techniques such as eDNA will be critical in furthering our understanding of NEP shark biology and the conservation status of specific populations. Such methods will also be important in assessing the viability and survival potential of these populations in the face of changing environments. The results of conservation genetic studies, once obtained, may not always translate into definitive conservation strategies; they do, however, provide an excellent starting point for management decisions (Avise and Hamerick, 2004). Yet poor communication, management constraints, lack of precedent, and politics may hinder the incorporation of genetic data into conservation and management plans. Issues with dissemination of data could be somewhat resolved by scientists playing a more active role in management, and by providing results of studies in 'plain

English', such that scientific knowledge is translated into management action (Kinnaird and O'Brien, 2001; Stinchcombe et al., 2002; Vernesi et al., 2008). Political constraints are considerably more difficult to overcome, because there are often disconnects between the measures that need to be taken and what can be feasibly and economically accomplished. As a result, conservation and management actions of endangered marine species are often done according to man made jurisdictional boundaries rather than natural population boundaries (Carr et al., 2003; Crowder et al., 2006). Yet success stories like the nascent recovery of the White Shark population off the coast of California (Burgess et al., 2014) and the rebound of the Tope Shark after the elimination of the gill-net fishery (Pondella and Allen, 2008) highlight both the effectiveness of well-applied shark management, and the potential value of additional powerful tools, such as conservation genetics, to the protection of elasmobranch populations across the globe.

ACKNOWLEDGEMENTS

We wish to thank the reviewers of this chapter, Dayv Lowry, Chris Chabot, and Lorenz Hauser, for thoughtful comments. We would also like to thank The Seattle Aquarium Management for supporting this effort.

REFERENCES

Avise, J.C., 1994. Molecular Markers, Natural History, and Evolution, second ed. Sunderland, MA, Sinauer Associates, Inc.

Avise, J., Hamerick, P., 2004. Conservation Genetics: Case Studies From Nature. Chapman & Hall, New York.

Barker, A.M., Nosal, A.P., Lewallen, E.A., Burton, R.S., 2015. Genetic structure of leopard shark (*Triakis semifasciata*) populations along the Pacific coast of North America. J. Exp. Mar. Biol. Ecol. 472, 151–157.

Barrera-García, A., O'Hara, T., Galván-Magaña, F., Méndez-Rodríguez, L.C., Castellini, J.M., Zenteno-Savín, T., 2012. Oxidative stress indicators and trace elements in the blue shark (*Prionace glauca*) off the east coast of the Mexican Pacific Ocean. Comp. Biochem. Physiol. C Toxicol. Pharmacol. 156, 59–66.

Bartl, S., 1998. What sharks can tell us about the evolution of MHC genes. Immunol. Rev. 166, 317–331.

Bartl, S., Nonaka, M., 2014. mhC molecules of cartilaginous fishes. In: Smith, S.L., Sim, R.B., Martin, F. (Eds.), Immunobiology of the Shark. CRC Press, Boca Raton, FL, p. 173.

Baum, J.K., Myers, R.A., 2004. Shifting baselines and the decline of pelagic sharks in the Gulf of Mexico. Ecol. Lett. 7 (2), 135–145.

Baum, J.K., Myers, R.A., Kehler, D.G., Worm, B., Harley, S.J., Doherty, P.A., 2003. Collapse and conservation of shark populations in the Northwest Atlantic. Science 299 (5605), 389–392.

Bentzen, P., Harris, A.S., Wright, J.M., 1991. Cloning of hypervariable minisatellite and simple sequence microsatellite repeats for DNA fingerprinting of important aquacultural species and salmonids and tilapia. In: Burke, T., Dolf, G., Jeffreys, A.L., Wolf, R.

(Eds.), DNA Fingerprinting: Approaches and Applications. Birkhauser-Verlag, Basel, Switzerland, pp. 243–262.

Blair, C., Campbell, C.R., Yoder, A.D., 2015. Assessing the utility of whole genome amplified DNA for next-generation molecular ecology. Mol. Ecol. Resour. 15 (5), 1079–1090.

Bossdorf, O., Richards, C.L., Pigliucci, M., 2008. Epigenetics for ecologists. Ecol. Lett. 11, 106–115.

Buonocore, F., Gerdol, M., 2016. Alternative adaptive immunity strategies: coelacanth, cod and shark immunity. Mol. Immunol. 69, 157–169.

Burgess, G.H., Bruce, B.D., Cailliet, G.M., Goldman, K.J., Grubbs, R.D., Lowe, C.G., et al., 2014. A re-evaluation of the size of the white shark (*Carcharodon carcharias*) population off California, USA. PLoS One 9 (6), e98078.

Byrne, R.J., Avise, J.C., 2012. Genetic mating system of the brown smoothhound shark (*Mustelus henlei*), including a literature review of multiple paternity in other elasmobranch species. Mar. Biol. 159, 749–756.

Cardeñosa, D., Hyde, J., Caballero, S., 2014. Genetic diversity and population structure of the pelagic thresher shark (*Alopias pelagicus*) in the Pacific Ocean: evidence for two evolutionarily significant units. PLoS One 9 (10), e110193. http://dx.doi.org/10.1371/journal.pone.0110193.

Carr, M.H., Neigel, J.E., Estes, J.A., Andelman, S., Warner, R.R., Largier, J., 2003. Comparing marine and terrestrial ecosystems: implications for the design of coastal marine reserves. Ecol. Appl. 13, S90–S107.

Cartamil, D., Wraith, J., Wegner, N.C., Kacev, D., Lam, C.H., Santana-Morales, O., Sosa-Nishizaki, O., Escobedo-Olvera, M., Kohin, S., Graham, J.B., 2016. Movements and distribution of juvenile common thresher sharks *Alopias vulpinus* in Pacific coast waters of the USA and Mexico. Mar. Ecol. Prog. Ser. 548, 153–163.

Castro, A.L., Stewart, B.S., Wilson, S.G., Hueter, R.E., Meekan, M.G., Motta, P.J., Bowen, B.W., Karl, S.A., 2007. Population genetic structure of Earth's largest fish, the whale shark (*Rhincodon typus*). Mol. Ecol. 16 (24), 5183–5192.

Chabot, C.L., 2015. Microsatellite loci confirm a lack of population connectivity among globally distributed populations of the tope shark *Galeorhinus galeus* (Triakidae). J. Fish Biol. 87 (2), 371–385.

Chabot, C.L., Allen, L.G., 2009. Global population structure of the tope (*Galeorhinus galeus*) inferred by mitochondrial control region sequence data. Mol. Ecol. 18 (3), 545–552.

Chabot, C.L., Haggin, B.M., 2014. Frequency of multiple paternity varies between two populations of brown smoothhound shark, *Mustelus henlei*. Mar. Biol. 161, 797–804.

Chabot, C.L., Espinoza, M., Mascareñas-Osorio, I., Rocha-Olivares, A., 2015. The effect of biogeographic and phylogeographic barriers on gene flow in the brown smoothhound shark, *Mustelus henlei*, in the northeastern Pacific. Ecol. Evol. 5 (8), 1585–1600.

Chapman, D.D., Firchau, B., Shivji, M.S., 2008. Parthenogenesis in a large-bodied requiem shark, the blacktip *Carcharhinus limbatus*. J. Fish Biol. 73 (6), 1473–1477.

Chapman, D.D., Feldheim, K.A., Papastamatiou, Y.P., Hueter, R.E., 2015. There and back again: a review of residency and return migrations in sharks, with implications for population structure and management. Ann. Rev. Mar. Sci. 7, 547–570.

Chen, Y.H., Lin, S.J., Lin, M.W., et al., 2002. Microsatellite polymorphism in promoter of heme oxygenase-1 gene is associated with susceptibility to coronary artery disease in type 2 diabetic patients. Hum. Genet. 111, 1–8.

Chen, M., Zou, M., Yang, L., He, S., 2012. Basal jawed vertebrate phylogenomics using transcriptomic data from Solexa sequencing. PLoS One 7, e36256.

Ciezarek, A.G., Dunning, L.T., Jones, C.S., Noble, L.R., Humble, E., Stefanni, S.S., Savolainen, V., 2016. Substitutions in the glycogenin-1 gene are associated with the evolution of endothermy in sharks and tunas. Genome Biol. Evol. 8, 3011–3021.

Corrigan, S., Kacev, D., Werry, J., 2015. A case of genetic polyandry in the shortfin mako *Isurus oxyrinchus*. J. Fish Biol. 87, 794–798.

Criscitiello, M.F., Saltis, M., Flajnik, M.F., 2006. An evolutionarily mobile antigen receptor variable region gene: doubly rearranging NAR-TcR genes in sharks. Proc. Natl. Acad. Sci. U.S.A. 103 (13), 5036–5041.

Crow, J.F., 1999. Hardy, Weinberg and language impediments. Genetics 152 (3), 821–825.

Crowder, L.B., Osherenko, G., Young, O.R., et al., 2006. Resolving mismatches in U.S. ocean governance. Science 313, 617–618.

Daly-Engel, T.S., Grubbs, R.D., Feldheim, K.A., Bowen, B.W., Toonen, R.J., 2010. Is multiple mating beneficial or unavoidable? Low multiple paternity and genetic diversity in the shortspine spurdog (*Squalus mitsukurii*). Mar. Ecol. Prog. Ser. 403, 255–267.

Daly-Engel, T.S., Seraphin, K.D., Holland, K.N., Coffey, J.P., Nance, H.A., Toonen, R.J., Bowen, B.W., 2012. Global phylogeography with mixed-marker analysis reveals male-mediated dispersal in the endangered scalloped hammerhead shark (*Sphyrna lewini*). PLoS One 7 (1), e29986.

Daly-Engel, T.S., Grubbs R.D., Pereira E., Larson, S., and Reece, J., (2017) Turning back the molecular clock: evolution and speciation in the Bluntnose Sixgill Shark (*Hexanchus griseus*) (in preparation).

Deck, C.A., Bockus, A.B., Seibel, B.A., Walsh, P.J., 2016. Effects of short-term hyper-and hypo-osmotic exposure on the osmoregulatory strategy of unfed North Pacific spiny dogfish (*Squalus suckleyi*). Comp. Biochem. Physiol. A Mol. Integr. Physiol. 193, 29–35.

Domeier, M.L., Nasby-Lucas, N., 2013. Two-year migration of adult female white sharks (*Carcharodon carcharias*) reveals widely separated nursery areas and conservation concerns. Anim. Biotelem. 1, 1.

Dowd, W.W., Harris, B.N., Cech, J.J., Kültz, D., 2010. Proteomic and physiological responses of leopard sharks (*Triakis semifasciata*) to salinity change. J. Exp. Biol. 213, 210–224.

Dudgeon, C.L., Broderick, D., Ovenden, J.R., 2009. IUCN classification zones concord with, but underestimate, the population genetic structure of the zebra shark *Stegostoma fasciatum* in the Indo-West Pacific. Mol. Ecol. 18 (2), 248–261.

Dulvy, N.K., Fowler, S.L., Musick, J.A., Cavanagh, R.D., Kyne, P.M., Harrison, L.R., Carlson, J.K., Davidson, L.N., Fordham, S.V., Francis, M.P., Pollock, C.M., 2014. Extinction risk and conservation of the world's sharks and rays. eLife 3, e00590.

Duncan, K.M., Martin, A.P., Bowen, B.W., De Couet, H.G., 2006. Global phylogeography of the scalloped hammerhead shark (*Sphyrna lewini*). Mol. Ecol. 15 (8), 2239–2251.

Ebert, D.A., White, W.T., Goldman, K.J., Compagno, L.J.V., Daly-Engel, T.S., Ward, R.D., 2010. Resurrection and redescription of *Squalus suckleyi* (Girard, 1854) from the North Pacific, with comments on the *Squalus acanthias* subgroup (Squaliformes: Squalidae). Zootaxa 2612, 22–40.

Edwards, S.V., Hedrick, P.W., 1998. Evolution and ecology of MHC molecules: from genomics to sexual selection. Trends Ecol. Evol. 13, 305–311.

Etter, P.D., Bassham, S., Hohenlohe, P.A., Johnson, E.A., Cresko, W.A., 2011. SNP discovery and genotyping for evolutionary genetics using RAD sequencing. Methods Mol. Biol. 772, 157–178.

Feldheim, K.A., Gruber, S.H., Ashley, M.V., 2004. Reconstruction of parental microsatellite genotypes reveals female polyandry and philopatry in the lemon shark, *Negaprion brevirostris*. Evolution 58, 2332–2342.

Feldheim, K.A., Gruber, S.H., Dibattista, J.D., Babcock, E.A., Kessel, S.T., Hendry, A.P., Pikitch, E.K., Ashley, M.V., Chapman, D.D., 2014. Two decades of genetic profiling yields first evidence of natal philopatry and long-term fidelity to parturition sites in sharks. Mol. Ecol. 23 (1), 110–117.

Fitzpatrick, J.L., Kempster, R.M., Daly-Engel, T.S., Collin, S.P., Evans, J.P., 2012. Assessing the potential for post-copulatory sexual selection in elasmobranchs. J. Fish Biol. 80, 1141–1158.

Frankham, R., 2010. Where are we in conservation genetics and where do we need to go? Conserv. Genet. 11 (2), 661–663.

Frick, L.H., Reina, R.D., Walker, T.I., 2010. Stress related physiological changes and post-release survival of Port Jackson sharks (*Heterodontus portusjacksoni*) and gummy sharks (*Mustelus antarcticus*) following gill-net and longline capture in captivity. J. Exp. Mar. Biol. Ecol. 385 (1), 29–37.

Gallagher, A.J., Serafy, J.E., Cooke, S.J., Hammerschlag, N., 2014. Physiological stress response, reflex impairment, and survival of five sympatric shark species following experimental capture and release. Mar. Ecol. Prog. Ser. 496, 207–218.

Garcia de Leaniz, C., Fleming, I.A., Einum, S., Verspoor, E., Jordan, W.C., Consuegra, S., Aubin-Horth, N., Lajus, D., Letcher, B.H., Youngson, A.F., Webb, J.H., 2007. A critical review of adaptive genetic variation in Atlantic salmon: implications for conservation. Biol. Rev. 82 (2), 173–211.

Goshima, M., Sekiguchi, R., Matsushita, M., Nonaka, M., 2016. The complement system of elasmobranches revealed by liver transcriptome analysis of a hammerhead shark, *Sphyrna zygaena*. Dev. Comp. Immunol. 61, 13–24.

Gubili, C., Robinson, C.E., Cliff, G., Wintner, S.P., de Sabata, E., De Innocentiis, S., Canese, S., Sims, D.W., Martin, A.P., Noble, L.R., Jones, C.S., 2015. DNA from historical and trophy samples provides insights into white shark population origins and genetic diversity. Endanger. Species Res. 27 (3), 233–241.

Guffey, S.C., Goss, G.G., 2014. Time course of the acute response of the North Pacific spiny dogfish shark (*Squalus suckleyi*) to low salinity. Comp. Biochem. Physiol. A Mol. Integr. Physiol. 171, 9–15.

Hauser, L., 2009. The molecular ecology of dogfish sharks. In: Gallucci, V., McFarlane, G., Bargmann, G. (Eds.), Biology and Management of Dogfish Sharks. American Fisheries Society, Bethesda, MD, pp. 229–252.

Heberer, C., Aalbers, S.A., Bernal, D., Kohin, S., DiFiore, B., Sepulveda, C.A., 2010. Insights into catch-and-release survivorship and stress-induced blood biochemistry of common thresher sharks (*Alopias vulpinus*) captured in the southern California recreational fishery. Fish. Res. 106, 495–500.

Hebert, P.D., Cywinska, A., Ball, S.L., 2003. Biological identifications through DNA barcodes. Proc. R. Soc. Lond. B Biol. Sci. 270 (1512), 313–321.

Hedrick, P.W., Miller, P.S., 1992. Conservation genetics: techniques and fundamentals. Ecol. Appl. 2 (1), 30–46.

Heist, E.J., Musick, J.A., Graves, J.E., 1996. Genetic population structure of the shortfin mako (*Isurus oxyrinchus*) inferred from restriction fragment length polymorphism analysis of mitochondrial DNA. Can. J. Fish. Aquat. Sci. 53, 583–588.

Hoelzel, A.R., 2001. Shark fishing in fin soup. Conserv. Genet. 2 (1), 69–72.

Hoelzel, A.R., Shivji, M.S., Magnussen, J., Francis, M.P., 2006. Low worldwide genetic diversity in the basking shark (*Cetorhinus maximus*). Biol. Lett. 2, 639–642.

Hoffmayer, E.R., Hendon, J.M., Parsons, G.R., Driggers Iii, W.B., Campbell, M.D., 2015. A comparison of single and multiple stressor protocols to assess acute stress in a coastal shark species, *Rhizoprionodon terraenovae*. Fish Physiol. Biochem. 41, 1253–1260.

Hooker, S.K., Gerber, L.R., 2004. Marine reserves as a tool for ecosystem-based management: the potential importance of megafauna. Bioscience 54, 27–39.

Jhaveri, P., Papastamatiou, Y.P., German, D.P., 2015. Digestive enzyme activities in the guts of bonnethead sharks (*Sphyrna tiburo*) provide insight into their digestive strategy and evidence for microbial digestion in their hindguts. Comp. Biochem. Physiol. A Mol. Integr. Physiol. 189, 76–83.

Jones, E., Chu, W.C., Ayele, M., Ho, J., Bruggeman, E., Yourstone, K., Rafalski, A., Smith, O.S., McMullen, M.D., Bezawada, C., Warren, J., Babayev, J., Basu, S., Smith, S., 2009. Development of single nucleotide polymorphism (SNP) markers for use in commercial maize (*Zea mays* L.) germplasm. Mol. Breed. 24 (2), 165–176.

Jorgensen, S.J., Reeb, C.A., Chapple, T.K., Anderson, S., Perle, C., Van Sommeran, S.R., Fritz-Cope, C., Brown, A.C., Klimley, A.P., Block, B.A., 2009. Philopatry and migration of Pacific white sharks. Proc. R. Soc. Lond. B Biol. Sci.rspb 20091155.

Karl, S.A., Castro, A.L., Garla, R.C., 2012. Population genetics of the nurse shark (*Ginglymostoma cirratum*) in the western Atlantic. Mar. Biol. 159 (3), 489–498.

Kashiwagi, T., Maxwell, E.A., Marshall, A.D., Christensen, A.B., 2015. Evaluating manta ray mucus as an alternative DNA source for population genetics study: underwater-sampling, dry-storage and PCR success. PeerJ 3, e1188.

Keeney, D.B., Heupel, M.R., Hueter, R.E., Heist, E.J., 2005. Microsatellite and mitochondrial DNA analyses of the genetic structure of blacktip shark (*Carcharhinus limbatus*) nurseries in the northwestern Atlantic, Gulf of Mexico, and Caribbean Sea. Mol. Ecol. 14 (7), 1911–1923.

Kelly, R.P., Port, J.A., Yamahara, K.M., Crowder, L.B., 2014. Using environmental DNA to census marine fishes in a large mesocosm. PLoS One 9 (1), e86175.

Kinnaird, M.F., O'Brien, T.G., 2001. Who's scratching whom? Reply to Whitten et al. Conserv. Biol. 15, 1459–1460.

Krishnamurthy, P.K., Francis, R.A., 2012. A critical review on the utility of DNA barcoding in biodiversity conservation. Biodivers. Conserv. 21 (8), 1901–1919.

Krück, N.C., Innes, D.I., Ovenden, J.R., 2013. New SNPs for population genetic analysis reveal possible cryptic speciation of eastern Australian sea mullet (*Mugil cephalus*). Mol. Ecol. Resour. 13 (4), 715–725.

Kulski, J.K., Shiina, T., Anzai, T., Kohara, S., Inoko, H., 2002. Comparative genomic analysis of the MHC: the evolution of class I duplication blocks, diversity and complexity from shark to man. Immunol. Rev. 190, 95–122.

Larson, S., Christiansen, J., Griffing, D., Ashe, J., Lowry, D., Andrews, K., 2011. Relatedness and polyandry of sixgill sharks, *Hexanchus griseus*, in an urban estuary. Conserv. Genet. 12, 679–690.

Larson, S., Farrer, D., Lowry, D., Ebert, D.A., 2015. Preliminary observations of population genetics and relatedness of the broadnose sevengill shark, *Notorynchus cepedianus*, in two Northeast Pacific Estuaries. PLoS One 10, e0129278.

Lewallen, E.A., Anderson, T.W., Bohonak, A.J., 2007. Genetic structure of leopard shark (*Triakis semifasciata*) populations in California waters. Mar. Biol. 152 (3), 599–609.

Li, Y.C., Korol, A.B., Fahima, T., Nevo, E., 2004. Microsatellites within genes: structure, function, and evolution. Mol. Biol. Evol. 21 (6), 991–1007.

Li, C., Hofreiter, M., Straube, N., Corrigan, S., Naylor, G.J.P., 2013. Capturing protein-coding genes across highly divergent species. Biotechniques 54, 321–326.

Lieber, L., Berrow, S., Johnston, E., Hall, G., Hall, J., Gubili, C., Sims, D.W., Jones, C.S., Noble, L.R., 2013. Mucus: aiding elasmobranch conservation through non-invasive genetic sampling. Endanger. Species Res. 21, 215–222.

Lighten, J., Incarnato, D., Ward, B.J., van Oosterhout, C., Bradbury, I., Hanson, M., Bentzen, P., 2016. Adaptive phenotypic response to climate enabled by epigenetics in a K-strategy species, the fish *Leucoraja ocellata* (Rajidae). R. Soc. Open Sci. 3 (10), 160299.

Magnussen, J.E., Pikitch, E.K., Clarke, S.C., Nicholson, C., Hoelzel, A.R., Shivji, M.S., 2007. Genetic tracking of basking shark products in international trade. Anim. Conserv. 10 (2), 199–207.

Mariette, S., Le Corre, V., Austerlitz, F., Kremer, A., 2002. Sampling within the genome for measuring within-population diversity: trade-offs between markers. Mol. Ecol. 11 (7), 1145–1156.

Marshall, H., Field, L., Afiadata, A., Sepulveda, C., Skomal, G., Bernal, D., 2012. Hematological indicators of stress in longline-captured sharks. Comp. Biochem. Physiol. A Mol. Integr. Physiol. 162, 121–129.

Martin, A.P., 1999. Substitution rates of organelle and nuclear genes in sharks: implicating metabolic rate (again). Mol. Biol. Evol. 16 (7), 996–1002.

Martin, A.P., Naylor, G.J., Palumbi, S.R., 1992. Rates of mitochondrial DNA evolution in sharks are slow compared with mammals. Nature 357 (6374), 153.

McFarlane, G., King, G., Leask, K., Christensen, L.B., 2009. Assessment of Information Used to Develop a Recovery Potential Assessment of Basking Shark *Cetorhinus maximus* (Pacific Population) in Canada. DFO Canadian Science Advisory Secretariat. Research Document. 2008/.71. 98 p.

Milano, I., Babbucci, M., Cariani, A., Atanassova, M., Bekkevold, D., Carvalho, G.R., Espiñeira, M., Fiorentino, F., Garofalo, G., Geffen, A.J., Hansen, J., 2014. Outlier SNP markers reveal fine-scale genetic structuring across European hake populations (*Merluccius merluccius*). Mol. Ecol. 23 (1), 118–135.

Mondo, K., Hammerschlag, N., Basile, M., Pablo, J., Banack, S.A., Mash, D.C., 2012. Cyanobacterial neurotoxin β-N-methylamino-L-alanine (BMAA) in shark fins. Mar. Drugs 10, 509–520.

Morin, P.A., Luikart, G., Wayne, R.K., 2004. SNPs in ecology, evolution and conservation. Trends Ecol. Evol. 19 (4), 208–216.

Morin, P.A., Archer, F.I., Foote, A.D., Vilstrup, J., Allen, E.E., Wade, P., Durban, J., Parsons, K., Pitman, R., Li, L., Bouffard, P., 2010. Complete mitochondrial genome phylogeographic analysis of killer whales (*Orcinus orca*) indicates multiple species. Genome Res. 20 (7), 908–916.

Musick, J.A., Burgess, G., Cailliet, G., Camhi, M., Fordham, S., 2000. Management of sharks and their relatives (Elasmobranchii). Fisheries 25 (3), 9–13.

Nei, M., 1987. Molecular Evolutionary Genetics. Columbia University Press, New York.

Nielsen, E.E., Hemmer-Hansen, J.A.K.O.B., Larsen, P.F., Bekkevold, D., 2009. Population genomics of marine fishes: identifying adaptive variation in space and time. Mol. Ecol. 18 (15), 3128–3150.

Nosal, A.P., Lewallen, E.A., Burton, R.S., 2013. Multiple paternity in leopard shark (*Triakis semifasciata*) litters sampled from a predominantly female aggregation in La Jolla, California, USA. J. Exp. Mar. Biol. Ecol. 446, 110–114.

O'Bryhim, J.R., Parsons, E.C.M., Lance, S.L., 2017. Forensic species identification of elasmobranch products sold in Costa Rican markets. Fish. Res. 186, 144–150.

O'Brien, S.M., Gallucii, V.F., Hauser, L., 2013. Effects of species biology on the historical demography of sharks and their implications for likely consequences of contemporary climate change. Conserv. Genet. 14 (1), 125–144.

Ogden, R., Dawnay, N., McEwing, R., 2009. Wildlife DNA forensics—bridging the gap between conservation genetics and law enforcement. Endanger. Species Res. 9 (3), 179–195.

Ohta, Y., Flajnik, M.F., 2015. Coevolution of MHC genes (LMP/TAP/class Ia, NKT-class Ib, NKp30-B7H6): lessons from cold-blooded vertebrates. Immunol. Rev. 267, 6–15.

Ohta, Y., Okamura, K., McKinney, E.C., Bartl, S., Hashimoto, K., Flajnik, M.F., 2000. Primitive synteny of vertebrate major histocompatability complex class I and class II genes. Proc. Natl. Acad. Sci. 97, 4712–4717.

Okamura, K., Ototake, M., Nakanishi, T., Kurosawa, Y., Hashimoto, K., 1997. The most primitive vertebrates with jaws possess highly polymorphic MHC class I genes comparable to those of humans. Immunity 7, 777–790.

Opazo, J.C., Lee, A.P., Hoffmann, F.G., Toloza-Villalobos, J., Burmester, T., Venkatesh, B., Storz, J.F., 2015. Ancient duplications and expression divergence in the globin gene superfamily of vertebrates: insights from the elephant shark genome and transcriptome. Mol. Biol. Evol. 32, 1684–1694.

Ovenden, J.R., Kashiwagi, T., Broderick, D., Giles, J., Salini, J., 2009. The extent of population genetic subdivision differs among four co-distributed shark species in the Indo-Australian archipelago. BMC Evol. Biol. 9, 40.

Pardini, A.T., Jones, C.S., Noble, L.R., Kreiser, B., Malcolm, H., Bruce, B.D., Stevens, J.D., Cliff, G., Scholl, M.C., Francis, M., Duffy, C.A.J., Martin, A.P., 2001. Sex-biased dispersal of great white sharks. Nature 412, 139–140.

Park, L.K., Moran, P., 1994. Developments in molecular genetic techniques in fisheries. Rev. Fish Biol. Fish. 4 (3), 272–299.

Paulson, T.G., Wright, F.A., Parker, B.A., Russack, V., Wahl, G.M., 1996. Microsatellite instability correlates with reduced survival and poor disease prognosis in breast cancer. Cancer Res. 56 (17), 4021–4026.

Phillips, N.M., Chaplin, J.A., Peverell, S.C., Morgan, D.L., 2017. Contrasting population structures of three *Pristis* sawfishes with different patterns of habitat use. Mar. Freshw. Res. 68, 452–460.

Pilliod, D.S., Goldberg, C.S., Arkle, R.S., Waits, L.P., 2014. Factors influencing detection of eDNA from a stream-dwelling amphibian. Mol. Ecol. Resour. 14 (1), 109–116.

Pondella, D.J., Allen, L.G., 2008. The decline and recovery of four predatory fishes from the southern California Bight. Mar. Biol. 154, 307–313.

Port, J.A., O'Donnell, J.L., Romero-Maraccini, O.C., Leary, P.R., Litvin, S.Y., Nickols, K.J., Yamahara, K.M., Kelly, R.P., 2016. Assessing vertebrate biodiversity in a kelp forest ecosystem using environmental DNA. Mol. Ecol. 25, 527–541.

Portnoy, D.S., Puritz, J.B., Hollenbeck, C.M., Gelsleichter, J., Chapman, D., Gold, J.R., 2015. Selection and sex-biased dispersal in a coastal shark: the influence of philopatry on adaptive variation. Mol. Ecol. 24 (23), 5877–5885.

Pratt, H.L., Carrier, J.C., 2001. A review of elasmobranch reproductive behavior with a case study on the nurse shark, *Ginglymostoma cirratum*. Environ. Biol. Fishes 60, 157–188.

Primmer, C.R., 2009. From conservation genetics to conservation genomics. Ann. N. Y. Acad. Sci. 1162 (1), 357–368.

Prugnolle, F., de Meeus, T., 2002. Inferring sex-biased dispersal from population genetic tools: a review. Heredity 88, 161–165.

Richards, V.P., Suzuki, H., Stanhope, M.J., Shivji, M.S., 2013. Characterization of the heart transcriptome of the white shark (*Carcharodon carcharias*). BMC Genomics 14 (1), 697.

Robinson, D.P., Baverstock, W., Al-Jaru, A., Hyland, K., Khazanehdari, K.A., 2011. Annually recurring parthenogenesis in a zebra shark *Stegostoma fasciatum*. J. Fish Biol. 79 (5), 1376–1382.

Saville, K.J., Lindley, A.M., Maries, E.G., Carrier, J.C., Pratt, H.L.J., 2002. Multiple paternity in the nurse shark, *Ginglymostoma cirratum*. Environ. Biol. Fishes 63, 347–352.

Schmidt, J.V., Schmidt, C.L., Ozer, F., Ernst, R.E., Feldheim, K.A., Ashley, M.V., Levine, M., 2009. Low genetic differentiation across three major ocean populations of the whale shark, *Rhincodon typus*. PLoS One 4 (4), e4988.

Schrey, A.W., Heist, E.J., 2003. Microsatellite analysis of population structure in the shortfin mako (*Isurus oxyrinchus*). Can. J. Fish. Aquat. Sci. 60 (6), 670–675.

Selkoe, K.A., Toonen, R.J., 2006. Microsatellites for ecologists: a practical guide to using and evaluating microsatellite markers. Ecol. Lett. 9 (5), 615–629.

Shen, T., Lei, M., Wang, J., He, X., Li, X., Li, J., 2014. Molecular cloning, organization, expression and 3D structural analysis of the MHC class Ia gene in the whitespotted bamboo shark (*Chiloscyllium plagiosum*). Vet. Immunol. Immunopathol. 157, 111–118.

Sigsgaard, E.E., Nielsen, I.B., Back, S.S., Lorenzen, E.D., Robinson, D.P., Knudsen, S.W., Pedersen, M.W., Jaidah, M.A., Orlando, L., Willerslev, E., Moller, P.R., Thomsen, P.F., 2016. Population characteristics of a large whale shark aggregation inferred from seawater environmental DNA. Nat. Ecol. Evol. 1, 0004.

Simpfendorfer, C.A., Kyne, P.M., Noble, T.H., Goldsbury, J., Basiita, R.K., Lindsay, R., Shields, A., Perry, C., Jerry, D.R., 2016. Environmental DNA detects critically endangered largetooth sawfish in the wild. Endanger. Species Res. 30, 109–116.

Skomal, G.B., Mandelman, J.W., 2012. The physiological response to anthropogenic stressors in marine elasmobranch fishes: a review with a focus on the secondary response. Comp. Biochem. Physiol. A Mol. Integr. Physiol. 162, 146–155.

Squire Jr., J.L., 1990. Distribution and apparent abundance of the basking shark, *Cetorhinus maximus*, off the central and southern California coast, 1962–85. Mar. Fish. Rev. 52 (2), 8–11.

Stinchcombe, J., Moyle, L.C., Hudgens, B.R., Bloch, P.L., Chinnadurai, S., Morris, W.F., 2002. The influence of the academic conservation biology literature on endangered species recovery planning. Conserv. Ecol. 6 (2), 15.

Sugaya, K., Fukagawa, T., Matsumoto, K.I., Mita, K., Takahashi, E.I., Ando, A., Inoko, H., Ikemura, T., 1994. Three genes in the human MHC class III region near the junction with the class II: gene for receptor of advanced glycosylation end products, PBX2 homeobox gene and a notch homolog, human counterpart of mouse mammary tumor gene int-3. Genomics 23 (2), 408–419.

Taylor, H.R., Harris, W.E., 2012. An emergent science on the brink of irrelevance: a review of the past 8 years of DNA barcoding. Mol. Ecol. Resour. 12 (3), 377–388.

Thibodeau, S.N., Bren, G., Schaid, D., 1993. Microsatellite instability in cancer of the proximal colon. Science 260 (5109), 816–820.

Veilleux, H.D., Ryu, T., Donelson, J.M., Van Herwerden, L., Seridi, L., Ghosheh, Y., Berumen, M.L., Leggat, W., Ravasi, T., Munday, P.L., 2015. Molecular processes of transgenerational acclimation to a warming ocean. Nat. Clim. Chang. 5 (12), 1074–1078.

Venkatesh, B., Lee, A.P., Ravi, V., Maurya, A.K., Lian, M.M., Swann, J.B., Ohta, Y., Flajnik, M.F., Sutoh, Y., Kasahara, M., 2014. Elephant shark genome provides unique insights into gnathostome evolution. Nature 505, 174–179.

Verissimo, A., McDowell, J.R., Graves, J.E., 2010. Global population structure of the spiny dogfish *Squalus acanthias*, a temperate shark with an antitropical distribution. Mol. Ecol. 19, 1651–1662.

Vernesi, C., Bruford, M.W., Bertorelle, G., Pecchioli, E., Rizzoli, A., Hauffe, H.C., 2008. Where's the conservation in conservation genetics? Conserv. Biol. 22, 802–804.

Ward, R.D., Zemlak, T.S., Innes, B.H., Last, P.R., Hebert, P.D., 2005. DNA barcoding Australia's fish species. Philos. Trans. R. Soc. Lond. B 360 (1462), 1847–1857.

Ward, R.D., Holmes, B.H., White, W.T., Last, P.R., 2008. DNA barcoding Australasian chondrichthyans: results and potential uses in conservation. Mar. Freshw. Res. 59 (1), 57–71.

Whitney, N.M., Pratt, H.L., Carrier, J.C., 2004. Group courtship, mating behaviour and siphon sac function in the whitetip reef shark, *Triaenodon obesus*. Anim. Behav. 68, 1435–1442.

Wilcox, T.M., McKelvey, K.S., Young, M.K., Jane, S.F., Lowe, W.H., Whiteley, A.R., Schwartz, M.K., 2013. Robust detection of rare species using environmental DNA: the importance of primer specificity. PLoS One 8 (3), e59520.

Wosnick, N., Bornatowski, H., Ferraz, C., Afonso, A., Rangel, B.S., Hazin, F.H.V., Freire, C.A., 2017. Talking to the dead: using post-mortem data in the assessment of stress in tiger sharks (*Galeocerdo cuvier*) (Péron and Lesueur, 1822). Fish Physiol. Biochem, 43 (1), 165–178.

Wright, S., 1943. Isolation by distance. Genetics 28 (2), 114.

Wright, J.M., Bentzen, P., 1995. Microsatellites: genetic markers for the future. In: Molecular Genetics in Fisheries. Springer, Netherlands, pp. 117–121.

CHAPTER FOUR

Diet Composition and Trophic Ecology of Northeast Pacific Ocean Sharks

Joseph J. Bizzarro*,[1], Aaron B. Carlisle†, Wade D. Smith‡, Enric Cortés§
*Moss Landing Marine Laboratories, Moss Landing, CA, United States
†Hopkins Marine Station of Stanford University, Pacific Grove, CA, United States
‡University of British Columbia, Institute for the Oceans and Fisheries, Vancouver, BC, Canada
§National Marine Fisheries Service, Southeast Fisheries Science Center, Panama City Laboratory, FL, United States
[1]Corresponding author: e-mail address: jbizzarro@mlml.calstate.edu

A salmon shark (*Lamna ditropis*) captures a pink salmon (*Onchorhynchus gorbuscha*) in Prince William Sound, Alaska. Photo courtesy of Scot Anderson, Point Reyes National Seashore, Inverness, California.

Contents

Advances in Marine Biology, Volume 77
ISSN 0065-2881
http://dx.doi.org/10.1016/bs.amb.2017.06.001

Abstract

Although there is a general perception of sharks as large pelagic, apex predators, most sharks are smaller, meso- and upper-trophic level predators that are associated with the seafloor. Among 73 shark species documented in the eastern North Pacific (ENP), less than half reach maximum lengths >200 cm, and 78% occur in demersal or benthic regions of the continental shelf or slope. Most small (≤200 cm) species (e.g., houndsharks) and demersal, nearshore juveniles of larger species (e.g., requiem sharks) consume small teleosts and decapod crustaceans, whereas large species in pelagic coastal and oceanic environments feed on large teleosts and squids. Several large, pelagic apex predator species occur in the ENP, but the largest species (i.e., Basking Shark, Whale Shark) consume zooplankton or small nekton. Size-based dietary variability is substantial for many species, and segregation of juvenile and adult foraging habitats also is common (e.g., Horn Shark, Shortfin Mako). Temporal dietary differences are most pronounced for temperate, nearshore species with wide size ranges, and least pronounced for smaller species in extreme latitudes and deep-water regions. Sympatric sharks often occupy various trophic positions, with resource overlap differing by space and time and some sharks serving as prey to other species. Most coastal species remain in the same general region over time and feed opportunistically on variable prey inputs (e.g., season migrations, spawning, or recruitment events), whereas pelagic, oceanic species actively seek hot spots of prey abundance that are spatiotemporally variable. The influence of sharks on ecosystem structure and regulation has been downplayed compared to that of large teleosts species with higher per capita consumption rates (e.g., tunas, billfishes). However, sharks also exert indirect influences on prey populations by causing behavioural changes that may result in restricted ranges and reduced fitness. Except for food web modelling efforts in Alaskan waters, the trophic impacts of sharks are poorly incorporated into current ecosystem approaches to fisheries management in the NEP.

1. INTRODUCTION

There has been a historical fascination among the general public and scientists alike with the predatory aspect of sharks. The typical perception of sharks is that of large, pelagic, apex predator species that significantly influence the population dynamics of their prey. Most sharks, however, actually are smaller (<2 m), mesotrophic level predators that are associated with continental shelf and slope habitats. As such, many shark species are important prey items for larger sharks, teleosts, and marine mammals and therefore transfer energy between lower- and upper-trophic levels (Wetherbee et al., 2012). Rather than adhering to a generalized, static construct that encapsulates a singular predatory role, the trophic ecology of sharks is quite complicated and dynamic. Indeed, the trophic impact of sharks varies greatly among taxa and ecosystems and also can vary considerably within a species across ontogeny (Ebert, 2002; Kim et al., 2012; Lowry et al., 2007), seasons (Andrews et al., 2010; Carlisle et al., 2011; Ketchen, 1986), or even the diel cycle (e.g., Klimley et al., 1988; Nelson and Johnson, 1970; Sciarrotta and Nelson, 1977). As scientists move away from general diet inventories to more complex ecological and hypothesis-driven research, we are beginning to expose misconceptions about the predatory role of sharks and better understand their true function in marine ecosystems.

Seventy-three shark species are documented to occur between equatorial regions of the northeast Pacific Ocean (NEP) and the Bering Sea (Table 1). Fifty-two per cent of these species reach maximum lengths <2 m, and 78% occur in benthic or demersal regions of the continental shelf or slope. These species occur in a wide range of depths and habitats, with diversity peaking in temperate waters and declining substantially with increasing latitude (Table 1). Owing to the substantial differences in taxonomy, size, and distribution of NEP sharks, their trophic relationships are highly variable. In addition, the trophic ecology of NEP sharks is poorly known compared to those of some other regions (e.g., western North Atlantic, Australia, and South Africa), with only a few well-studied species (e.g., North Pacific Spiny Dogfish, *Squalus suckleyi*; Blue Shark, *Prionace glauca*; and Leopard Shark, *Triakis semifasciata*).

In this chapter, we synthesize and review the literature concerning trophic relationships, trophodynamics, and ecological impacts of sharks in the NEP, focusing on those species that range from the southern border of Mexico to the Bering Sea, and extend to abyssal regions of the seafloor

Table 1 Maximum Length, Depth Range, Distribution, and Habitat of 73 Shark Species That Occur in the Northeast Pacific (NEP), Arranged Phylogenetically

Order	Family	Genus and Species	Common Name	Max. Length (cm)	Min. Depth (m)	Max. Depth (m)	Distribution	Habitat
Heterodontiformes	Heterodontidae	*Heterodontus francisci*	Horn Shark	96	0	200	Tm	BN
		H. mexicanus	Mexican Horn Shark	70	0	67	TmTr	BN
		H. quoyi	Galapagos Bullhead Shark	107	0	40	Tr	BN
Orectolobiformes	Ginglymostomatidae	*Ginglymostoma unami*	Nurse Shark	208	0	55	TmTr	BN
	Rhincodontidae	*Rhincodon typus*	Whale Shark	1880	0	1700	TmTr	PNO
Lamniformes	Odontaspididae	*Odontaspis ferox*	Ragged-Tooth Shark	450	10	883	TmTr	DNO
	Mitsukurinidae	*Mitsukurina owstoni*	Goblin Shark	550	0	1300	TmTr	DPNO
	Pseudocarcharhinidae	*Pseudocarcharias kamoharai*	Crocodile Shark	122	0	590	TmTr	DPO
	Megachasmidae	*Megachasma pelagios*	Megamouth Shark	550	0	600	TmTr	PNO
	Alopiidae	*Alopias pelagicus*	Pelagic Thresher	365	0	700	TmTr	PO
		A. superciliosus	Bigeye Thresher	480	0	723	TmTr	PO
		A. vulpinis	Common Thresher Shark	636	0	572	BTmTr	PNO

	Cetorhinidae	*Cetorhinus maximus*	Basking Shark	980	0	1264	BTm	PNO
	Lamnidae	*Carcharodon carcharias*	White Shark	640	0	1280	BTmTr	DPNO
		Isurus oxyrinchus	Shortfin Mako	396	0	740	TmTr	PNO
		I. paucus	Longfin Mako	430	0	760	TmTr	PO
		Lamna ditropis	Salmon Shark	310	0	1864	BTm	PNO
Carcharhiniformes	Scyliorhinidae	*Apristurus brunneus*	Brown Catshark	71	33	1306	BTm	BDPN
		A. kampae	Longnose Catshark	65	180	1888	Tm	BDN
		A. nasutus	Largenose Catshark	59	400	925	TmTr	BDN
		A. stenseni	Panama Ghost Catshark	23	915	975	Tr	BDN
		Bythaeulurus giddingsi	Galapagos Catshark	45	428	562	Tr	BDN
		Cephaloscyllium ventriosum	Swell Shark	110	5	457	TmTr	BDN
		Cephalurus cephalus	Lollipop Catshark	30	155	927	TmTr	BDN
		Galeus piperatus	Peppered Catshark	36	130	1326	Tm	BDN
		Parmaturus xaniurus	Filetail Catshark	61	88	1250	Tm	BDPN
	Triakidae	*Galeorhinus galeus*	Soupfin Shark	195	2	471	TmTr	DPN
		Mustelus albipinnis	Whitemargin Fin Smoothhound	118	30	280	Tm	DN

Continued

Table 1 Maximum Length, Depth Range, Distribution, and Habitat of 73 Shark Species That Occur in the Northeast Pacific (NEP), Arranged Phylogenetically—cont'd

Order	Family	Genus and Species	Common Name	Max. Length (cm)	Min. Depth (m)	Max. Depth (m)	Distribution	Habitat
		M. californicus	Grey Smoothhound	125	2	247	TmTr	DN
		M. dorsalis	Sharptooth Smoothhound	77	20	200	TmTr	DN
		M. henlei	Brown Smoothhound	106	2	281	Tm	DN
		M. lunulatus	Sicklefin Smoothhound	175	9	200	TmTr	DN
		M. mento	Speckled Smoothhound	130	16	50	TmTr	DN
		Triakis acutipinna	Sharpfin Houndshark	102	?	?	Tr	DN
		T. maculata	Spotted Houndshark	240	?	?	Tr	DN
		T. semifasciata	Leopard Shark	210	2	156	TmTr	DN
	Carcharinidae	*Carcharhinus albimarginatus*	Silvertip Shark	285	0	800	Tr	DPN
		C. altimus	Bignose Shark	285	12	803	TmTr	DPN
		C. brachyurus	Narrowtooth Shark	294	0	360	TmTr	DPN
		C. cerdale	Pacific Smalltail Shark	140	5	40	Tr	DPN

	C. falciformis	Silky Shark	330	0	500	TmTr	PNO
	C. galapagensis	Galapagos Shark	300	0	285	TmTr	DPNO
	C. leucas	Bull Shark	340	0	152	TmTr	DN
	C. limbatus	Blacktip Shark	255	0	160	TmTr	DPN
	C. longimanus	Oceanic Whitetip Shark	395	0	230	TmTr	PO
	C. obscurus	Dusky Shark	390	0	573	TmTr	DPN
	C. porosus	Smalltail Shark	150	5	32	TmTr	DN
	Galeocerdo cuvier	Tiger Shark	740	0	920	TmTr	DPN
	Nasolamia velox	Whitenose Shark	156	15	192	TmTr	DN
	Negaprion brevirostris	Lemon Shark	340	0	92	TmTr	BDN
	Prionace glauca	Blue Shark	383	0	812	BTmTr	PO
	Rhizoprionodon longurio	Pacific Sharpnose Shark	129	0	100	TmTr	DN
	Triaenodon obesus	Whitetip Reef Shark	213	0	330	Tr	DN

Continued

Table 1 Maximum Length, Depth Range, Distribution, and Habitat of 73 Shark Species That Occur in the Northeast Pacific (NEP), Arranged Phylogenetically—cont'd

Order	Family	Genus and Species	Common Name	Max. Length (cm)	Min. Depth (m)	Max. Depth (m)	Distribution	Habitat
	Sphyrnidae	*Sphyrna corona*	Scalloped Bonnethead	92	?	?	Tr	DN
		S. lewini	Scalloped Hammerhead	430	0	980	TmTr	DPNO
		S. media	Scoophead	150	0	100	Tr	DN
		S. mokarran	Great Hammerhead	610	0	300	Tr	PNO
		S. tiburo	Bonnethead	150	0	80	TmTr	DN
		S. zygaena	Smooth Hammerhead	400	0	200	TmTr	PNO
Hexanchiformes	Chlamydoselachidae	*Chlamydoselachus anguineus*	Frill Shark	196	0	1570	TmTr	DPN
	Hexanchidae	*Hexanchus griseus*	Bluntnose Sixgill Shark	550	0	2500	BTmTr	DN
		Notorynchus cepedianus	Broadnose Sevengill Shark	296	0	136	Tm	DN
Echinorhiniformes	Echinorhinidae	*Echinorhinus cookei*	Prickly Shark	450	4	1100	TmTr	DN
Squaliformes	Squalidae	*Squalus suckleyi*	North Pacific Spiny dogfish	130	0	1460	BTm	DPN
	Centrophoridae	*Centrophorus squamosus*	Leafscale Gulper Shark	164	145	2400	Tr	BDPO

	Etmopteridae	*Centroscyllium nigrum*	Combtooth Dogfish	52	269	1212	TmTr	DN
		Etmopterus benchleyi	Ninja Lanternshark	52	836	1443	Tr	BDN
	Somniosidae	*Somniosus pacificus*	Pacific Sleeper Shark	700	0	2205	BTm	BDN
		Zameus squamulosus	Velvet Dogfish	84	0	1450	TmTr	BDPNO
	Dalatiidae	*Euprotomicrus bispinatus*	Pygmy Shark	31	0	1800	TmTr	PO
		Isistius brasiliensis	Cookiecutter Shark	50	0	3500	TmTr	PO
Squatiniformes	Squatinidae	*Squatina armata*	Chilean Angel Shark	150	30	50	TmTr	BN
		S. californica	Pacific Angel Shark	152	3	205	Tm	BN

NEP = Equator to Bering Sea, including abyssal plains and offshore oceanic regions east of the Hawaiian Archipelago; Max. Length = total length; (NEP) Distribution: B = boreal, Tm = temperate, Tr = tropical; Habitat: B = benthic, D = demersal, N = nearshore, P = pelagic, O = oceanic, ? = unknown.

and overlying oceanic habitats east of the Hawaiian Archipelago. We also conduct a metaanalysis of the standardized diet compositions and trophic levels of NEP sharks to advance our understanding of shark trophic ecology in this region. We then utilize these results to define trophic guilds and examine sources of dietary variability (e.g., taxon, total length, depth, and habitat) in NEP sharks. Many sharks are highly mobile, and these movements are often driven by, or coordinated with, foraging behaviour. We review the growing body of literature that documents the behaviours and movement patterns of NEP sharks as they relate to foraging across a range of spatial and temporal scales. We then take a more holistic look at the ecological roles of NEP sharks, including their direct and indirect predatory impacts, and the role of sharks as prey. Finally, we examine current approaches to fisheries management in the NEP that incorporate trophic information on sharks, and suggest possibilities for further exploration and development.

2. STANDARDIZED DIET COMPOSITIONS AND TROPHIC LEVELS

2.1 Background and Methods

Quantitative diet composition estimates provide a basis for understanding a species' prey spectrum and its overall trophic ecology. Diet composition estimates also are important inputs for descriptive and comparative trophic analyses, and for the calculation of additional measures, such as dietary diversity and trophic level indices. Standardized diet composition and trophic level estimates are especially important for large marine predators that may influence ecosystem dynamics but typically are poorly surveyed for food habit information. A series of peer-reviewed manuscripts have been published to address this need.

The focal group of standardized diet composition and trophic level manuscripts has varied during the last 20 years, with most of the work centred on elasmobranchs. The objectives, analyses, and applications of the research also expanded over time. The initial work was conducted on marine mammals (Pauly et al., 1998) and provided the general objectives and protocol for subsequent works. Cortés (1999) modified the marine mammal prey categories of Pauly et al. (1998) to be more suitable for sharks and replaced the original rank approach to diet composition estimates with a synthesis of information obtained from single (e.g., % Frequency of Occurrence, %FO; % Number, %N; and % Weight, %W) or, preferably, compound indices (e.g., Index of

Relative Importance, %IRI). Ebert and Bizzarro (2007) modified the composition of generalized prey categories from Cortés (1999) to better match the diet of skates, increased the number of prey categories, and introduced statistical analysis to compare diet composition among species. A complementary study on rays (Jacobsen and Bennett, 2013) largely subscribed to the methods and approach of Ebert and Bizzarro (2007). The most recent work, published on US Pacific Coast groundfishes, incorporated a variety of additional goals and analyses to apply trophic results to pressing issues in habitat-based fisheries management (Bizzarro et al., 2017). The information presented in this section represents an updated version of Cortés (1999) for NEP sharks, whereas Section 3 uses the generated diet composition and trophic level estimates to investigate sources of dietary variability.

The methods used here for standardized diet composition and trophic level calculations were synthesized from Cortés (1999) and Bizzarro et al. (2017). The primary objectives of this analysis are to: (1) calculate standardized diet compositions and trophic levels of NEP sharks; (2) determine the primary sources of dietary variation among NEP sharks; (3) determine whether diet composition and trophic level differ among taxonomic or functional groups; and (4) distinguish groups of NEP sharks with similar diet compositions (i.e., trophic guilds). Diet composition information was only gathered from studies that sampled sharks within the NEP, as defined in Section 1, and ranged from 1921 to 2008. Prey categories were adopted from Cortés (1999). Because weight (or volume) is a proxy for energetic importance, and therefore the most appropriate input for trophic level calculations, weight estimates were extracted from literature sources whenever possible and represented as %W. When weight estimates were not available, the order of preference for dietary metrics was compound index, numerical, and finally occurrence. Bray–Curtis dissimilarity was used as the basis of matrix calculations for multivariate analyses. PERMANOVA (with permutation tests of multivariate group dispersions, Anderson, 2006) was used to determine the best combination of response variables to explain the observed dietary variability in the dataset (after Bizzarro et al., 2017). Variables (and their factors) included: family, order, maximum documented total length, median depth of occurrence, geographic distribution (boreal, temperate, tropical), and the vertical (benthic, demersal, pelagic) and horizontal (nearshore, oceanic) dimensions of habitat. Trophic level comparisons were made using one-way ANOVA. Trophic guilds were characterized using hierarchical agglomerative cluster analysis (after Bizzarro et al., 2017).

2.2 Results

Standardized diet compositions and trophic levels were calculated for 30 shark species occurring in the NEP. Eighty-one sources of quantitative diet composition information yielded a total of 18,561 samples from individual sharks, with nearly 75% of these obtained from North Pacific Spiny Dogfish (Table 2). Fourteen species were represented by <100 samples, including six species with <20 samples that were not included in comparative analyses (Table 2). Diet composition information for the White Shark, *Carcharodon carcharias*, was available from eight studies but because of the difficulty in obtaining samples, the total number of sampled individuals was low.

The studied group of NEP sharks was primarily composed of piscivores (Table 2). Fish occurred in the diets of 93.3% of species with mean and median values of 39.8% and 37.0%, respectively. Four species had relatively robust sample sizes and diets consisting of >70% fish: Shortfin Mako, *Isurus oxyrinchus*; Salmon Shark, *Lamna ditropis*; Tope Shark, *Galeorhinus galeus*; Pacific Sharpnose Shark, *Rhizoprionodon longurio*. Cephalopods (%FO = 76.7, %Mean = 18.5) and crustaceans (%FO = 73.3, %Mean = 24.7) were of similar secondary importance in the diets of NEP sharks, but only two species, the Blue Shark and Blacktip Shark, *Carcharhinus limbatus*, consumed both taxa in large quantities (Table 2). Pelagic sharks such as hammerheads (*Sphyrna* spp.), carcharhinids (e.g., Silky Shark, *Carcharhinus falciformis*; Oceanic Whitetip, *C. longimanus*, Blue Shark), and lamnids (e.g., Common Thresher, *Alopias vulpinis*; Shortfin Mako, Salmon Shark) consumed cephalopods in large proportions (Table 2). Crustaceans were substantial dietary components for small carcharhiniforms such as catsharks (Scyliorhinidae) and smoothhounds (Triakidae) (Table 2). Consumption of zooplankton and hard-shelled molluscs was rare, but these prey taxa were dominant in the diet of the Basking Shark, *Cetorhinus maximus*, and Horn Shark, *Heterodontus francisci*, respectively. By contrast, marine reptiles were not recorded in the diet of any species and sea birds were trivial prey items in the diets of only two species (Table 2). Marine mammals were dominant in the diet of NEP White Sharks and were reported in the diets of three other large, coastal species (Table 2). Chondrichthyans were consumed by a third of all NEP shark species, mainly in low quantities; however, they were particularly abundant in the diet of cow sharks (Bluntnose Sixgill Shark, *Hexanchus griseus*; Broadnose Sevengill Shark, *Notorynchus cepedianus*).

NEP shark species comprised a mixture of tertiary (trophic level, TL > 4) and secondary (TL > 3) consumers (Table 2). Estimated trophic levels ranged from 3.20 (Horn Shark, Basking Shark) to 4.82 (White Shark), with

Table 2 Standardized Diet Compositions and Trophic Levels of NEP Sharks, Arranged Phylogenetically

Order	Family	Genus and Species	*n*	N	FISH	CEPH	MOL	CR	INV	ZOO	BIR	REP	MAM	CHON	PL	Trophic Level
Heterodontiformes	Heterodontidae	*Heterodontus francisci*	1	97	0.0	0.1	71.3	27.0	0.2	0.0	0.0	0.0	0.0	0.0	1.4	3.20
Lamniformes	Megachasmidae	*Megachasma pelagios*	1	1	0.0	0.0	0.0	50.0	50.0	0.0	0.0	0.0	0.0	0.0	0.0	3.51
	Alopiidae	*Alopias superciliosus*	1	1	100.0	0.0	0.0	0.0	0.0	0.0	0.0	0.0	0.0	0.0	0.0	4.24
		A. vulpinis	4	740	58.9	40.7	0.0	0.4	0.0	0.0	0.0	0.0	0.0	0.0	0.0	4.22
	Cetorhinidae	*Cetorhinus maximus*	1	9	0.0	0.0	0.0	0.0	0.0	100.0	0.0	0.0	0.0	0.0	0.0	3.20
	Lamnidae	*Carcharodon carcharias*	8	35	18.1	0.0	0.0	1.0	0.0	0.0	0.0	0.0	69.8	11.1	0.0	4.82
		Isurus oxyrinchus	4	150	73.0	25.9	0.0	1.1	0.0	0.0	0.0	0.0	0.0	0.0	0.0	4.22
		Lamna ditropis	1	102	82.1	16.8	0.0	0.0	0.0	0.0	0.0	0.0	0.0	1.1	0.0	4.24
Carcharhiniformes	Scyliorhinidae	*Apristurus brunneus*	2	138	19.7	3.3	0.1	74.9	1.0	1.0	0.0	0.0	0.0	0.0	0.0	3.68
		Cephalurus cephalus	1	9	20.0	0.0	0.0	70.0	10.0	0.0	0.0	0.0	0.0	0.0	0.0	3.66
		Parmaturus xaniurus	1	70	6.7	0.1	0.0	91.0	0.0	2.2	0.0	0.0	0.0	0.0	0.0	3.56

Continued

Table 2 Standardized Diet Compositions and Trophic Levels of NEP Sharks, Arranged Phylogenetically—cont'd

Order	Family	Genus and Species	*n*	N	FISH	CEPH	MOL	CR	INV	ZOO	BIR	REP	MAM	CHON	PL	Trophic Level
	Triakidae	*Galeorhinus galeus*	1	170	90.3	7.5	0.0	0.0	0.0	0.0	0.0	0.0	0.0	2.2	0.0	4.25
		Mustelus californicus	2	119	0.0	0.2	0.1	99.3	0.3	0.0	0.0	0.0	0.0	0.0	0.0	3.52
		M. dorsalis	1	192	31.9	0.2	0.0	67.8	0.0	0.0	0.0	0.0	0.0	0.0	0.0	3.75
		M. henlei	2	72	19.1	8.1	0.0	72.7	0.0	0.0	0.0	0.0	0.0	0.0	0.0	3.71
		M. lunulatus	2	230	21.3	0.0	2.1	76.7	0.0	0.0	0.0	0.0	0.0	0.0	0.0	3.66
		Triakis semifasciata	6	870	32.9	1.0	3.2	36.4	25.0	0.0	0.0	0.0	0.0	0.0	1.3	3.73
	Carcharinidae	*Carcharhinus falciformis*	3	151	25.2	71.8	0.0	3.0	0.0	0.0	0.0	0.0	0.0	0.0	0.0	4.19
		C. limbatus	1	35	53.8	30.8	0.0	15.4	0.0	0.0	0.0	0.0	0.0	0.0	0.0	4.12
		C. longimanus	1	38	46.2	50.0	0.0	0.0	0.0	0.0	0.0	0.0	0.0	3.8	0.0	4.24
		Prionace glauca	7	412	41.1	39.5	0.2	15.9	2.2	0.0	0.2	0.0	0.0	0.6	0.2	4.09
		Rhizoprionodon longurio	2	104	98.9	0.3	0.0	0.6	0.0	0.0	0.1	0.0	0.0	0.0	0.0	4.24
	Sphyrnidae	*Sphyrna lewini*	4	211	44.6	53.2	0.0	2.2	0.0	0.0	0.0	0.0	0.0	0.0	0.0	4.20
		S. zygaena	2	15	58.9	39.7	0.0	0.0	0.0	0.0	0.0	0.0	0.0	1.4	0.0	4.23

Hexanchiformes	Hexanchidae	*Hexanchus griseus*	3	55	47.8	0.0	0.0	0.0	0.0	0.0	0.0	0.0	8.3	43.9	0.0	4.48
		Notorynchus cepedianus	3	458	32.9	0.0	0.0	0.3	0.0	0.0	0.0	0.0	15.2	51.6	0.0	4.57
Squaliformes	Squalidae	*Squalus suckleyi*	6	13,820	47.7	4.3	0.2	25.3	7.1	12.2	0.0	0.0	0.0	3.1	0.1	3.88
	Somniosidae	*Somniosus pacificus*	7	219	46.6	49.7	0.2	0.1	0.0	0.0	0.0	0.0	3.4	0.0	0.0	4.24
	Dalatiidae	*Isistius brasiliensis*	2	11	27.3	63.6	0.0	9.1	0.0	0.0	0.0	0.0	0.0	0.0	0.0	4.15
Squatiniformes	Squatinidae	*Squatina californica*	1	27	48.2	48.2	0.0	0.0	0.0	0.0	0.0	0.0	0.0	0.0	3.6	4.14

n = number of studies used in calculations, N = total number on nonempty stomachs among studies. Prey categories include: telost fishes (FISH), cephalopods (CEPH), other molluscs (MOL), crustaceans (CR), other invertebrates (INV), marine birds (BIR), marine reptiles (REP), marine mammals (MAM), chondrichthyan fishes (CHON), and plants (PL). A complete list of references used in diet composition calculations is available by contacting the senior author.

a mean (MTL) of 4.00 and a median of 4.14. Trophic levels generally increased with maximum documented size for NEP sharks, with the notable exceptions of the diminutive Cookiecutter Shark (*Isistius brasiliensis*, TL=4.15) and massive Basking Shark (TL=3.20). Significant ordinal differences in MTL were determined ($F=4.73$, $P=0.013$), and a Tukey HSD test indicated that hexanchiforms (4.53) and lamniforms (4.38) had significantly higher MTLs than carcharhiniforms (3.92). The MTL of squaliforms (4.06) did not differ significantly from any other orders. MTL also differed significantly among shark families ($F=10.24$, $P<0.001$). Hexanchids (4.53), lamnids (4.43), and carcharhinids (4.17) had significantly higher MTL levels than scyliorhinids (3.62) or triakids (3.77).

3. DIETARY VARIABILITY

Family-level differences were the primary source of dietary variability among the studied NEP sharks, indicating a strong positive association between taxonomic relatedness and dietary similarity. The 24 analysed species comprised 11 families, however, limiting replication and reducing the inherent variability of the data set. Three other variables, maximum total length, vertical habitat, and order, displayed significant differences among groups (categorical variables) or along a gradient (continuous variable) (Table 3). The best-fit model only contained a single variable (Family), although the addition of vertical habitat approached significance at the 0.05 level ($r^2 = 0.088$, $P = 0.062$). Variance differed significantly among horizontal habitat groups, but was relatively homogeneous for all other categorical variables.

Species were grouped into trophic guilds based on similar diet compositions (Fig. 1). An agglomerative cluster coefficient of 0.96 indicated a high degree of clustering throughout the dissimilarity range. Additionally, pairwise species dissimilarities and cophenetic distances associated with the cluster dendrogram were strongly correlated (0.77), signifying that the dendrogram is an appropriate summary of diet data for NEP shark species.

Four significant trophic guilds were detected, encompassing 21 of the 24 analysed species (>20 stomach samples). Three species (Horn Shark, Leopard Shark, and North Pacific Spiny Dogfish) formed a nonsignificant group with diets that included crustaceans and differentially high proportions of relatively scarce prey items (i.e., other molluscs, other invertebrates, and zooplankton). A mixture of smoothhounds and catsharks represented a significant guild with diets dominated by crustaceans (mean=80.4%) (Fig. 1).

Table 3 PERMANOVA Models of Diet Composition Among Several Taxonomic and Functional Group Variables for NEP Sharks With Sample Sizes >20 (See Table 2)

Model(s)	Variable(s)	df	*F*	r^2	*P*
Final model	Family	10	2.83	0.685	0.001
	Residuals	*13*		*0.315*	
Independent variables	Family	10	2.83	0.685	0.001
	Maximum total length	1	6.72	0.234	0.001
	Vertical habitat	5	2.42	0.402	0.009
	Order	5	2.13	0.371	0.016
	Horizontal habitat	2	1.91	0.153	0.089*
	Geographic distribution	3	0.91	0.120	0.507
	Median depth	1	0.76	0.033	0.523

Degrees of freedom (df), *F*-statistic, amount of variability explained (r^2), *P*-value, and results of dispersion analysis (*$P < 0.05$) are included.

White, Bluntnose Sixgill, and Broadnose Sevengill Sharks formed a significant guild of apex predators that consumed relatively high proportions of marine mammal and chondrichthyan prey (Fig. 1). Another small guild identified highly piscivorous species (Pacific Sharpnose Shark, Tope Shark, and Salmon Shark). The diet composition of the largest significant trophic guild, consisting of nine species, was dominated by a mixture of fishes and cephalopods. Although not significant, the two most dissimilar clusters grouped the species assemblage by size, indicating that smaller species mainly ate crustaceans, whereas the diets of larger species consisted primarily of teleost and chondrichthyan fishes, cephalopods, and marine mammals.

Other sources of dietary variability have been reported among NEP sharks but cannot be captured in a generalized metaanalysis at the species level. Ontogenetic dietary changes are almost universal in fishes (Gerking, 1994) and have been reported for several NEP species (e.g., Ebert, 2002; Kim et al., 2012). In many cases, however, these studies actually report differences in size and not maturity. Spatiotemporal dietary variability is commonly reported in sharks (Wetherbee et al., 2012) and is presented by a synthesis of behavioural and ecological studies in the ensuing sections of this chapter. Sex-based dietary differences have been reported in NEP sharks (e.g., Tope shark, Ripley, 1946) but typically represent differences in size or reflect spatial differences in prey resources that result from sexual segregation (McCord and Campana, 2003; Sims, 2006).

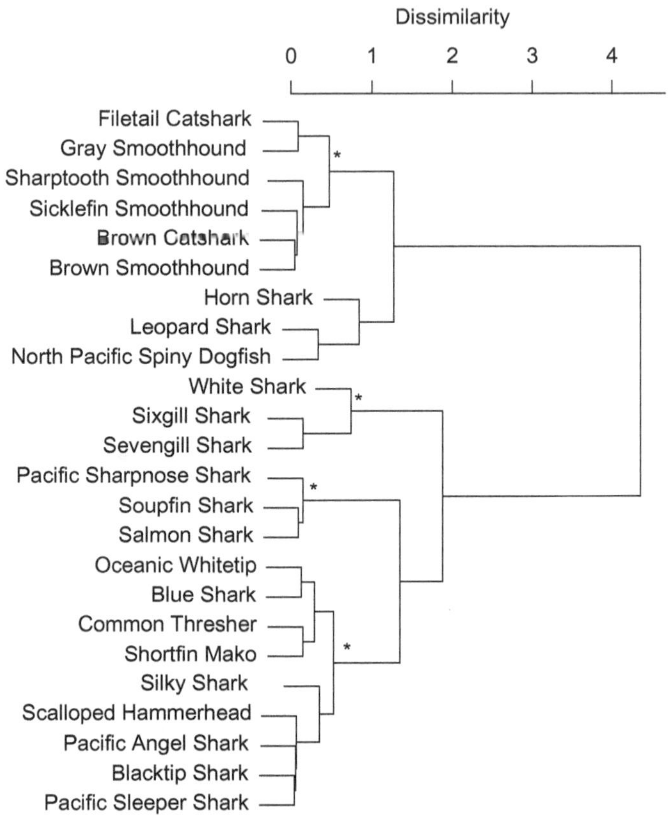

Fig. 1 Hierarchical agglomerative cluster analysis of standardized diet composition among 24 NEP shark species with sample sizes >20. Calculations were performed using Ward's Minimum Variance technique, and a Bray–Curtis dissimilarity matrix. *Significant cluster ($P < 0.05$).

4. FORAGING MOVEMENTS, PERIODICITY, AND BEHAVIOUR

The distribution of sharks is influenced by many abiotic and biotic factors, including physiology (Bernal and Lowe, 2015), environmental conditions (Schlaff et al., 2014), habitat availability (Simpfendorfer and Heupel, 2012), and the distribution of prey (Heithaus et al., 2002). Yet from a trophic perspective, the relevant aspects of their distribution patterns are those related to direct food–web interactions. The ecological roles that sharks play in maintaining community structure and function continue to be debated, but sharks, particularly upper-trophic level or apex predator species, may play an important role in marine food webs through both direct (i.e., predation) and indirect (e.g., competition) effects (Heithaus et al., 2010; Navia et al., 2012). Sharks also may have indirect, nonconsumptive effects by altering the behaviour of potential prey and competitors (Lima and Dill, 1990), but it is their role as consumers that directly impacts their fitness and characterizes their function in marine food webs. To play a direct role in a particular food web, the distributions of shark and other predators and prey organisms must overlap (i.e., sympatric predator–prey distributions). Hence, the trophic ecology and overall ecological role of sharks is inherently linked to their distributional patterns. In this section, we provide a review of the periodicity, behaviours, and movements of sharks in the NEP that have been directly attributed to foraging. Examples span a breadth of spatial and temporal scales, ranging from short-term, small-scale movements to larger-scale seasonal movements and migrations.

4.1 Foraging Movements, Migrations, and Aggregations: Tidal Patterns

In shallow coastal environments, especially bays and estuaries, tidal cycles directly influence habitat availability and a range of physiochemical factors, such as currents, turbidity, salinity, temperature, and dissolved oxygen. As a result, tidal cycles play a fundamental role in structuring the distribution of sharks and their prey in these environments (Carlisle and Starr, 2010). In the NEP, tidal foraging movements have been reported in triakid sharks in the bays and estuaries of Central and Northern California. Leopard Sharks and Brown Smoothhounds (*Mustelus henlei*) have been observed moving onto intertidal mudflats of Tomales Bay, CA, to forage during the flood tide (Ackerman et al., 2000; Campos et al., 2009). In Elkhorn Slough, CA, Leopard Sharks exhibit two different patterns of tidal movement that are habitat dependent (Carlisle and Starr, 2009, 2010). When in areas with extensive intertidal mudflats, sharks move onto low intertidal mudflats during the flood tide, where they feed on abundant invertebrate prey until forced back to adjacent subtidal habitats during the ebb tide. When in deeper channel habitats, sharks move with the tides, up and down the length of the Slough. By moving with the tides, sharks are able to use the currents to forage across a larger area and feed upon a prey field that is more dispersed relative to that of tidal mudflats.

4.2 Foraging Movements, Migrations, and Aggregations: Diel Patterns

Whereas tidal patterns commonly influence the foraging behaviour of sharks in shallow environments, diel patterns are more often observed in coastal and pelagic sharks. These diel patterns may reflect temporal changes in activity level, depth distribution, or geographic location associated with foraging. Sharks can use their enhanced sensory capabilities to effectively forage in low light conditions, whether at night or at depth (Hueter et al., 2004; Tricas, 1982). In the kelp forests and rocky reefs of the NEP, Horn Sharks, and Swell Sharks (*Cephaloscyllium ventriosum*) are largely quiescent during the day and increase activity levels at night when they primarily feed, at times using electroreception to locate prey (Nelson and Johnson, 1970; Tricas, 1982). Pacific Angel Shark activity levels also generally are higher during the night, with peaks of activity during crepuscular periods, reflecting individuals actively moving and foraging throughout relatively small home ranges (Standora and Nelson, 1977).

Vertical movements allow sharks to expand their potential prey base or access different prey resources and are particularly common in pelagic sharks. In marine ecosystems, and pelagic ones in particular, diel changes in distribution or behaviours of predators frequently are associated with diel changes in prey distribution, such as vertical migration of organisms associated with the deep scattering layer (Neilson and Perry, 1990). In Southern California, juvenile White Sharks, juvenile Shortfin Makos, and Common Thresher Sharks primarily reside in shallow waters during the night and dive more during the day to feed upon pelagic, demersal, or benthic prey (Cartamil et al., 2010; Sepulveda et al., 2004; Weng et al., 2007b). A Megamouth Shark (*Megachasma pelagios*) tracked in Southern California exhibited diel vertical migration, remaining at shallow depths during the night, and diving to deeper depths during the daytime, potentially following a particular light level or vertically migrating prey (Nelson et al., 1997). Salmon Sharks in Alaska exhibit a seasonal shift in depth distribution related to changes in the distribution and availability of prey, although they are generally more shallowly distributed during the night than during the day throughout the year (Carlisle et al., 2011). Trawl surveys less commonly caught North Pacific Spiny Dogfish on or near the bottom during the night, suggesting that they move up into the water column during the night to feed (Orlov et al., 2011). Blue Sharks in the Channel Islands of California were tracked moving inshore to feed on abundant squid during the night, and then back offshore to deeper water habitats during the day (Sciarrotta and Nelson, 1977). In oceanic habitats of the North Pacific Transition Zone, Blue Sharks feed primarily during the daytime, presumably diving to feed upon deep- and mid-water squid (Kubodera et al., 2007). In the Gulf of California, Scalloped Hammerhead Sharks (*Sphyrna lewini*) form large daytime aggregations around seamounts, but disperse during the night to feed in pelagic waters (Klimley et al., 1988).

Benthic and demersal sharks also exhibit diel shifts in depth distribution that are related to foraging. Off central California, Prickly Sharks (*Echinorhinus cookei*) largely are restricted to the deeper waters of the Monterey Bay Submarine Canyon during daylight hours, but ascend to the head of the Canyon at night to feed in the water column (Dawson and Starr, 2009). Similarly, Pacific Sleeper Sharks (*Somniosus pacificus*) in the Gulf of Alaska and Bluntnose Sixgill Sharks in Puget Sound, WA, remain at depth during the day and ascend to shallow daytime habitats to feed during the night (Andrews et al., 2009; Hulbert et al., 2006; Orlov and Baitalyuk, 2014).

Some species show a high degree of plasticity in diel feeding periodicity across their range and even within habitats. Leopard Sharks in Elkhorn Slough and Tomales Bay, CA, though strongly influenced by tides, appear to have some degree of diel periodicity to their feeding behaviour (Ackerman et al., 2000; Carlisle and Starr, 2010), with increased activity levels indicative of increased foraging at night. Kao (2000) reported that Leopard Sharks in Elkhorn Slough mainly feed on fish at night, suggesting that the increased rates of movement may reflect a more active foraging behaviour relative to one focused on relatively slow or immobile benthic invertebrates (Carlisle and Starr, 2010). Leopard Sharks in Southern California (Catalina Island, La Jolla) primarily exhibit a diel periodicity to their foraging behaviour, where they leave daytime aggregation regions in shallow sandy beach habitats to forage in deeper waters at night (Manley, 1995; Nosal et al., 2013, 2014). Grey Smoothhounds (*Mustelus californicus*) appear to forage in subtidal mudflat and eelgrass habitats in Bolsa Chica lagoon more often during the night (Espinoza et al., 2011).

4.3 Larger-Scale Movements and Migrations

Although diel or tidal patterns influence the foraging behaviour of sharks over shorter time frames, larger scale, seasonal changes in environmental conditions, prey distribution, or the maturation mating/parturition needs of an individual can influence the broad scale distribution or migratory patterns of sharks. Because most highly migratory sharks are upper-trophic level predators (see Section 2.2), these movements can influence food web dynamics and energy transfer among distinct ecosystems. The distribution of Bluntnose Sixgill Sharks in Puget Sound, WA, shifts in response to seasonal changes in prey availability, moving north during the winter and spring and south during the summer and fall (Andrews et al., 2010). Broadnose Sevengill Sharks seasonally move from coastal habitats into bays and estuaries in the NEP to feed and reproduce (Ebert, 1989; Williams et al., 2012). North Pacific Spiny Dogfish exhibit seasonal foraging migrations, moving northward in spring and summer as water temperatures increase, and then back south during the fall as water temperatures drop (Ketchen, 1986; McFarlane and King, 2003; Orlov et al., 2011). Blue Sharks off Catalina Island, CA, shift from the diel onshore–offshore movement pattern described earlier during the spring and summer to remaining offshore during late summer and fall in response to seasonal changes in prey availability (Sciarrotta and Nelson, 1977).

Seasonal migrations are common in migratory species, and in the NEP the large-scale migratory patterns of lamnid sharks have been linked to complementary shifts in prey availability, environmental conditions, and reproduction (Block et al., 2011). During the summer and fall, Salmon Sharks aggregate in coastal habitats in Alaska to feed upon large numbers of salmon gathering at the mouths of their natal rivers prior to spawning (Hulbert et al., 2005). While some Salmon Sharks remain throughout the year in productive high latitude environments (Carlisle et al., 2011), other sharks migrate from their foraging habitats off Alaska and British Columbia, Canada, to feed in the productive waters off California during the spring, spending little time in less productive pelagic habitats during their migrations (Weng et al., 2008). In pelagic habitats of the North Pacific Transition Zone and Subarctic Gyre, highly correlated catch rates suggest that Salmon Shark may follow salmon schools (Nagasawa, 1998). Juvenile Salmon Shark in the California Current primarily occur offshore and feed on pelagic prey, but occasionally move inshore, potentially in response to shifts in prey availability, to feed upon more coastal prey (Carlisle et al., 2015).

Large White Sharks move between coastal aggregation sites, where they feed intensively on pinnipeds during the fall and winter, and oligotrophic oceanic habitats, though males and females exhibit different migratory patterns (Jorgensen et al., 2009; Weng et al., 2007a). Males exhibit a regular annual migration, where they feed at coastal aggregations sites in the fall and winter, move offshore for ~6–8 months, and then return the following fall or winter. Mature females have a less regular migratory pattern where they move offshore for longer periods, as long as ~1.5 years, before returning to coastal foraging habitats in the fall and winter (Chapple et al., 2016; Domeier and Nasby-Lucas, 2013; Jorgensen et al., 2009; Weng et al., 2007a). Both sexes appear to use the period of feeding on energy-rich marine mammals in coastal habitats to provision for their long offshore migrations (Del Raye et al., 2013), with some evidence suggesting that their rate of feeding decreases when offshore, at least in males (Carlisle et al., 2012). The diet composition of White Shark in offshore habitats is largely unknown, but stable isotope data indicate that they feed on oceanic prey, such as pelagic squid and teleosts, that are isotopically distinct from coastal prey (Carlisle et al., 2012).

4.4 Feeding Behaviours of NEP Sharks

Sharks in the NEP exhibit a wide range of feeding behaviours that allow them to capture and consume prey. Leopard Sharks and Grey Smoothhounds use their rostra to excavate buried prey or use suction to extract invertebrates from burrows (Ackerman, 1971; Talent, 1976, 1982). For Leopard Sharks, the capacity to generate suction and, thus, the type of prey items that can be effectively foraged for, also changes over ontogeny as a consequence of allometric changes in cranial morphology (Lowry et al., 2007). Juvenile Horn Sharks have been observed foraging in feeding pits excavated by Bat Rays (*Myliobatis californica*) (Strong, 1989). Pacific Angel Sharks partially bury themselves in the sediment to ambush mobile fishes, with some individuals having preferred ambush locations that they use repeatedly (Fouts and Nelson, 1999). Leopard Sharks and North Pacific Spiny Dogfish feed together on schools of forage fish, swimming slowly through the schools with their mouths open (Russo, 1975). Broadnose Sevengill Sharks employ a variety of hunting strategies (Ebert, 1991). When hunting marine mammals and elasmobranchs, Sevengills often use stealth and ambush tactics in addition to hunting cooperatively with conspecifics. They also have been observed to spy hop, where they elevate their head above the water to locate prey (Ebert, 1991). Pacific Sleeper and Bluntnose Sixgill Sharks use stealth to ambush prey and are known to consume prey through suction feeding (McNeil et al., 2016), as evidenced by the presence of large, intact organisms in their stomachs (Ebert et al., 1987; Hulbert et al., 2006; Orlov and Moiseev, 1999; Sigler et al., 2006).

Pelagic sharks also exhibit a broad array of hunting behaviours. One of the most dramatic behaviours is exhibited by thresher sharks (Alopidae), which use the elongated upper lobe of their caudal fin as a whip to stun prey (Aalbers et al., 2010). This behaviour results in thresher sharks commonly being caught by their tails in longline fisheries (Strasburg, 1958). Blue Sharks in Southern California exhibit a variety of feeding behaviours when foraging on schools of forage fishes and squids, that vary based on prey characteristics. Blue Sharks sometimes slowly swim through a prey school with their mouths open while moving their heads from side to side. At other times they charge directly through the school, indiscriminately consuming large numbers of prey, or adopt a vertical posture, feeding on the bottom part of the prey school using broad slow sweeps of their tail to maintain their position in the water column (Tricas, 1979). White Sharks also use a variety of hunting strategies, ranging from actively chasing prey to using stealth to approach from below and then rapidly ascending to strike and disable or kill the prey, at times letting it bleed to death before consuming it (Klimley, 1994; Tricas and McCosker, 1984). White Sharks also use stealth to approach and attack prey at depth, initiating attacks from behind and below the prey (Skomal et al., 2015).

In pelagic habitats where prey is less abundant, sharks use a variety of strategies to attract prey. Oceanic Whitetip Sharks are believed to use their distinctive coloration, with the white tips of their fins resembling a school of small fish from a distance, to lure potential prey (Myrberg, 1990). Cookiecutter Sharks appear to use their bioluminescent collar to attract larger predators, which the shark then attacks to excise a plug of tissue (Widder, 1998). Similarly, the Megamouth Shark has a bioluminescent strip on its lower jaw that has been hypothesized to function as a lure to attract prey (Nakaya, 2001).

5. ECOLOGICAL ROLES

Though the influence of sharks can be critical in shaping population and community dynamics (Ferretti et al., 2010; Heithaus et al., 2008, 2010; Heupel et al., 2014), their ecological roles are complex and often plastic. Diets will shift across life spans, with sharks targeting different types of prey or trophic positions through ontogeny (see Section 3). Highly mobile species influence and connect communities and ecosystems across broad spatial and temporal scales as they traverse ocean basins (e.g., Blue Shark, Nakano, 1994; North Pacific Spiny Dogfish, McFarlane and King, 2003) or latitudinal ecoregions (e.g., Salmon shark, Weng et al., 2008). Although apex predators, White and Pacific Sleeper Sharks also operate as facultative scavengers, feeding on carcasses of marine mammals (Long and Jones, 1996, Sigler et al., 2006). Additionally, the ecological impacts of sharks extend beyond direct consumption of prey. Interspecific competition and indirect predatory effects influence behaviours, interactions, and community dynamics (Heithaus et al., 2010). Robust dietary analyses and improved tracking methods have enabled an assessment of marine food web dynamics and the inference of critical functional responses to changes in environmental conditions and anthropogenic influences across ecologically relevant spatial and temporal scales. In this section, we present a brief overview of the mechanisms by which sharks influence NEP food webs through their roles as predators and prey. We highlight the paucity of directed investigations and modelling efforts to evaluate the diverse ecological roles of sharks in the communities and ecosystems of this region in an attempt to stimulate future research.

5.1 Direct Predatory Impacts

Feeding presents a clear and direct, typically lethal effect on prey species that has the potential to generate community-level changes in biomass and trophic structure. Research into the predatory roles of sharks has been stimulated, in part, by global declines of many shark species (Dulvy et al., 2014; Stevens et al., 2000). As populations of these upper-trophic level predators are reduced how will the community respond? Dietary studies offer preliminary insight into the trophic position and types of prey that are consumed by a particular species (see Section 2). In the NEP, direct predatory impacts of sharks have received increasing consideration at the population level, particularly among scientists studying marine mammals that are at relatively low or declining abundance (Horning and Mellish, 2014; Tinker et al., 2016). Predation by sharks (e.g., Bluntnose Sixgill, Broadnose Sevengill, White Sharks, and perhaps Pacific Sleeper Sharks) could represent a significant source of mortality among a number of marine mammals (Ebert, 2002; Horning and Mellish, 2014; Long and Jones, 1996). Though Southern Sea Otters (*Enhydra lutris nereis*) are not reflected in the diets of White Sharks, an increase in fatal, nonconsumptive predation mortality over the past decade represents a major source of mortality for this recovering otter population (Tinker et al., 2016). Because sea otters generally are considered to be keystone species in kelp forest ecosystems, this shark-induced increase in mortality could have important cascading ecosystem-level effects. Quantifying the effects of direct predation, however, requires dietary information as well as estimates of daily ration, metabolic rate, and population estimates of both predator and prey (Williams et al., 2004). Estimates of prey consumption by NEP sharks (e.g., North Pacific Spiny Dogfish, Andrews and Harvey, 2013; Shortfin Mako, Vetter et al., 2008) have been undertaken but the data required to estimate predatory impacts rarely have been available for both predator and prey species. Nagasawa (1998) presents one exception. Data on abundance, daily ration, and metabolic rates were incorporated to estimate the potential predatory impact of Salmon Sharks on salmonid populations in the North Pacific, leading Nagasawa (1998) to suggest that direct predation could remove 13%–25% of the overall annual biomass of Pacific salmon.

Sharks function as both consumers and prey within marine communities. In Section 2.2, we highlight the occurrence of other chondrichthyans in the diets of selected NEP sharks. However, these are not the only predators of sharks. The egg cases of oviparous sharks (including Swell and Horn Sharks) are commonly preyed upon by gastropods (Cox and Koob, 1993). Some bony fishes occasionally prey on sharks, mainly consuming relatively small species or juveniles of larger species (e.g., Bargmann, 1977). Marine mammals are potential competitors for resources as well as prey and predators of sharks. For example, we have observed California Sea Lions (*Zalophus californianus*) consume Leopard, Smoothhound (*Mustelus* spp.), and Thresher (*Alopias* spp.) sharks in southern California. Blue Sharks also have been documented as prey of California Sea Lions (Lowry et al., 1991). North Pacific Spiny Dogfish are reported in the diets of Harbour Seals (*Phoca vitulina*; Bromaghin et al., 2013) and California (Weise and Harvey, 2008) and Stellar (*Eumetopias jubatus*; Trites et al., 2006) Sea Lions. Cartilaginous fishes, including Pacific Angel Sharks, Blue Sharks, and Brown Catsharks (*Apristurus brunneus*), may be important components of the diet of Northern Elephant Seals (*Mirounga angustirostris*; Condit and Le Boeuf, 1984). Adult White, Shortfin Mako, and Pacific Sleeper Sharks are not exempt from predation and may be consumed by transient Killer (*Orcinus orca*) and potentially by Sperm (*Physeter macrocephalus*) Whales (Ford et al., 2011; Pyle et al., 1999).

In the absence of the necessary biological data to quantify the predatory role of sharks, inferences on the community-level consequences of shark depletion can be made from time-series analyses of catch and research survey records (e.g., Ferretti et al., 2010; Myers et al., 2007). Overexploitation of upper-trophic level species such as sharks should alter top-down forcing mechanisms and transmit cascading effects that may be detected by changes in the relative abundance of lower-trophic levels. Inferences based on correlations between the biomass of predators and prey can be prone to oversimplification and overextension but are highly informative when examined using competing alternative hypotheses, knowledge of trends in fishery practices and regulations that influence observed patterns, and consideration of life history. Polovina et al. (2009) provide an example of this correlative approach that is relevant to the North Pacific. An examination of trends in abundance drawn from fishery-dependent catch data across 10 years indicated an annual decline in the landings of the top predators in the system

(3%–9%), including Blue Sharks (Polovina et al., 2009). The catch rates of four mid-trophic level fishes concurrently increased by 6%–18% each year, suggesting that the observed increase in abundance of the mid-trophic level species may resulted from diminished top-down predation pressure. In contrast to other upper-trophic level predators, such as tunas, the life history characteristics of many sharks (e.g., long life span and low fecundity) produce greater sensitivity to fishing pressure. Schindler et al. (2002) compared the potential population and community trophic level consequences of fishing mortality on Blue Sharks and Yellowfin Tuna (*Thunnus albacares*) in the Central Pacific Ocean. Although tuna predation rates were 4–5 times greater than those of Blue Sharks, relatively low levels of fishing on Blue Sharks produced population declines that generated significant changes in food web dynamics (Schindler et al., 2002). Increased fishing pressure and large-scale changes in environmental conditions underscore the need for and importance of an improved understanding of the trophic interactions and functional responses of sharks in the NEP.

Dynamic trophic mass-balance models (Ecopath with Ecosim [EwE] and Ecospace) have been widely applied to reconstruct historic levels of abundance and test hypotheses about the potential consequences of perturbations to a community as a result of environmental change, fishing pressure, or alternative management strategies (Christensen and Walters, 2004; Stevens et al., 2000). These models incorporate a range of data on the abundance, energetics, diets, and life history of species to simulate interspecific interactions under equilibrium conditions. EwE simulations that include sharks as functional groups have been developed in a number of investigations in the NEP to evaluate trophic structure and function under varying environmental regimes or levels of exploitation (e.g., Harvey et al., 2012; Heymans et al., 2007; Stevens et al., 2000; Surma and Pitcher, 2015; Tolimieri et al., 2013). Most studies grouped sharks into single, coarse predatory guilds comprising a shared trophic level. However, Kitchell et al. (2002) applied more refined groupings to directly explore the potential influence of sharks to act as keystone predators in the Central Pacific. Their models incorporated distinct categories for Blue Shark, 'large sharks' (Mako and White Sharks), and 'brown sharks' (carcharhinid species including Silky and Oceanic Whitetip Sharks), as well as other large predatory fishes. The abundance of mid-trophic level fishes and squid increased following declines of sharks and other top predators as a result of doubled fishing pressure. When fishing mortality was eliminated as a factor, the abundance of sharks and other top predators increased and the same mid-trophic level species declined. Results of these simulations suggest that the suite of top predators (i.e., tuna, billfish, and sharks) collectively exerted top-down control, but none functioned explicitly as a keystone predator because when one top predator was removed from the system, the remaining ones occupied the vacant niche (Kitchell et al., 2002). Notably, Kitchell et al. (2002) found that the extent of cannibalism displayed by large sharks could exert strong nonlinear effects on food web dynamics and may represent an important and often unaddressed consideration when assessing the predatory role of sharks.

The modelling efforts by Kitchell et al. (2002) and evaluations of empirical data by Polovina et al. (2009) highlight the redundancies and complexity of interactions that exist within marine food webs. Sharks have the capacity to exert significant top-down effects through direct lethal interactions but the extent of these effects varies among systems and species. As discussed in the following section, indirect effects of predators can also be an important mechanism that structures marine communities (Heithaus et al., 2008; Lima and Dill, 1990). Trophic models such as EwE can, to a limited extent, incorporate important nonlethal predatory effects, such as predator avoidance, into simulations (Kitchell et al., 2002).

5.2 Indirect Predatory Impacts

In addition to being a direct source of mortality to prey, predators also can act as a source of stress that can alter behavioural and morphological traits on ecological and evolutionary time scales (Dill et al., 2003; Lima and Dill, 1990; Pressier et al., 2005). To reduce the risk of predation, a species may employ a range of behavioural responses including forming aggregations, shifting habitat use, altering movement patterns, or increasing vigilance (Dill et al., 2003; Heithaus et al., 2008). By prompting behavioural changes that produce shifts in the feeding tactics and distribution patterns of a prey species, a predator can indirectly affect predation rates of prey, thereby influencing the densities and survivorship of nonprey organisms. Such indirect effects may be equal to or more important than direct consumptive predator–prey interactions to prey population size and structure and to community dynamics (Anholt and Werner, 1995; Heithaus et al., 2008). Investigations into indirect effects of predation are scarce for marine communities and these studies have primarily been conducted in tropical environments (e.g., Heithaus and Dill, 2006; Heithaus et al., 2007). Computer simulations present an excellent mechanism for exploring the trophic consequences of indirect predatory effects on scales for which empirical multispecies biological and behavioural data are unavailable. In a series of theoretical modelling experiments using species from the NEP, Frid et al. (2006, 2008) demonstrated that the risk of predation by Pacific Sleeper Sharks could significantly alter the diving behaviour and diet of Harbor Seals. Models simulating Harbor Seal response to predators indicated an underutilization of deeper water where the probability of encounters with Pacific Sleeper Sharks would be elevated (Frid et al., 2008). Without predation risk from sharks, Harbor Seals increased the depths of their dives and fed on larger more energy-rich species. Results of these simulations indicate that indirect predatory effects may represent a significant influence on trophic dynamics in the NEP that have the potential to produce or amplify trophic cascades (Frid et al., 2006, 2008; Heithaus et al., 2008).

6. APPLICATIONS OF NEP SHARK TROPHIC ECOLOGY FOR ASSESSMENT AND MANAGEMENT

Direct applications of trophic ecology for ecosystem assessment and fisheries management are absent in subtropical and tropical waters of the NEP, but under development in temperate and boreal regions. Shark management is rudimentary or nonexistent throughout Central America (Kyne et al., 2012; Rojas et al., 2000; Salas et al., 2011), and, although Mexico has adopted a national shark management plan (Sosa-Nishizaki et al., 2008), it has yet to pursue an ecosystem approach to fisheries management. In the past decade the United States and Canada, however, both have legislated that ecosystem-based fisheries management (EBFM) practices drive national ocean policies (Department of Fisheries and Oceans Canada, 2007; Lubchenco and Sutley, 2007). Canadian efforts are focused on defining and understanding ecoregions and managing human activities to preserve ecosystem health. Integrated ecosystem assessments represent the scientific basis for the implementation of EBFM in the United States (Levin et al., 2009).

Most contemporary ecosystem research in the NEP is directed at determining ecosystem structure and function, and predicting ecosystem dynamics resulting from anthropogenic (e.g., fishing) or environmental (e.g., climate change) impacts. Accurate trophic inputs are essential to the performance of food web and more complex ecosystem models (e.g., Atlantis) that are increasingly used to predict and evaluate ecosystem responses. In one recent study, potential impacts of climate change on NEP marine food webs and fisheries were modelled, and sharks appeared vulnerable to substantial population declines under moderate or substantial climate change scenarios

(Ainsworth et al., 2011). Scientific uncertainty about ecological interactions negatively impacts conservation and recovery strategies for predators, especially those that share prey resources (Marshall et al., 2016). For example, a global study on management of apex predators found that recovery, such as that exhibited by the White Shark in the NEP (Burgess et al., 2014; Stier et al., 2016) is rare, and that among the most important factors complicating predator recoveries was a lack of baseline trophic information (e.g., diet, resource limitation, and competition) (Stier et al., 2016).

Knowledge of food–web interactions is needed to better understand the mechanisms controlling the spatiotemporal dynamics of sharks and their prey. In terms of fisheries management, the concept of stock-specific Maximum Sustainable Yield (MSY) is challenged when changes in the abundance of one stock affect the abundance of other stocks. For example, management measures aimed at increasing the biomass of sharks will increase predation rates on stocks of their prey, which may decrease below MSY levels. It is therefore crucial to quantify the trophic interactions of sharks and their prey to establish sustainable fishing levels in an ecosystem context. This requires knowledge of both the abundance of shark populations and their prey, and of shark consumption rates, to build minimally realistic ecosystem models. EBFM is a common goal of the National Oceanographic and Atmospheric Administration in the United States, Fisheries and Oceans Canada, and other management bodies and Regional Fisheries Management Organizations around the world. For sharks, implementation of multispecies EBFM is severely hampered by the paucity of data compared to other data rich, commercially important fish stocks. Only with a serious investment in collection of basic trophic data will implementation of EBFM be feasible.

The standardized diet compositions and trophic levels presented in Section 2 can be directly applied to improve the accuracy of ecosystem assessments and the development of EBFM practices concerning sharks in the NEP. Mean trophic level is one of the most widely used indicators of marine health and has notably been used to investigate fishing impacts on ecosystem structure (Essington et al., 2006; Pauly et al., 1998). There is a need, however, to decouple assessments of ecosystem health from MTL estimates generated from catch data, which contain inherent biases (Branch et al., 2010). The taxon-specific MTL estimates calculated here from quantitative dietary studies can serve as inputs for the inclusion of sharks in ecosystem models of the NEP.

Functional groups (or guilds) typically are used in modelling exercises to reduce complexity. Because functional group designations are based on the assumption that all species within a group have similar characteristics, accurate assignments are critical for effective model development (Fausch et al., 1990; Stobutzki et al., 2001). The results of our trophic guild analysis (Section 2.2) facilitate the accurate grouping of a suite of NEP shark species into functional groups for assessment and management purposes. The overall and relative strength of our trophic guild assignments are indicated by their associated statistical probabilities, which can be used as a proxy for data quality. Most prior shark functional groupings are imprecise and could better reflect similarities in diet composition by incorporating the results of our guild assignments (e.g., Field et al., 2006; Tolimieri et al., 2013). Additionally, the data sources used to characterize NEP shark diets for current modelling efforts are dated and highly generalized (Dufault et al., 2009). Incorporation of standardized diet composition data and identification of primary sources of dietary variability could be used to update current trophic and ecosystem models to more accurately depict responses of NEP sharks to dynamic future scenarios.

ACKNOWLEDGEMENTS

We thank Kit Johnston (National Marine Fisheries Service, Southwest Fisheries Science Center—Fisheries Ecology Division) for obtaining obscure and unavailable diet composition literature, Alexi Orlov and Felipe Galván-Magaña for their helpful reviews, and Dayv Lowry and Shawn Larson (Editors, this volume) for their patience and assistance with the completion of this chapter.

REFERENCES

Aalbers, S.A., Bernal, D., Supulveda, C.A., 2010. The functional role of the caudal fin in the feeding ecology of the common thresher shark *Alopias vulpinus*. J. Fish Biol. 76, 1863–1868.

Ackerman, L.T., 1971. Contributions to the Biology of the Leopard Shark, *Triakis semifasciata* (Girard) in Elkhorn Slough, Monterey Bay. Sacramento State College, California.

Ackerman, J.T., Kondratieff, M.C., Matern, S.A., Cech, J.J., 2000. Tidal influence on spatial dynamics of leopard sharks, *Triakis semifasciata*, in Tomales Bay, California. Environ. Biol. Fishes 58, 33–43.

Ainsworth, C.H., Samhouri, J.F., Busch, D.S., Cheung, W.W.L., Dunne, J., Okey, T.A., 2011. Potential impacts of climate change on Northeast Pacific marine foodwebs and fisheries. ICES J. Mar. Sci. 68, 1217–1229.

Anderson, M.J., 2006. Distance-based test for homogeneity of multivariate dispersions. Biometrics 62, 245–253.

Andrews, K.S., Harvey, C.J., 2013. Ecosystem-level consequences of movement: seasonal variation in the trophic impact of a top predator. Mar. Ecol. Prog. Ser. 473, 247–260.

Andrews, K.S., Williams, G.D., Farrer, D., Tolimieri, N., Harvey, C.J., Bargmann, G.G., Levin, P.S., 2009. Diel activity patterns of sixgill sharks, *Hexanchus griseus*: the ups and downs of an apex predator. Anim. Behav. 78, 525–536.

Andrews, K.S., Williams, G.D., Levin, P.S., 2010. Seasonal and ontogenetic changes in movement patterns of sixgill sharks. PLoS One 5. e12549.

Anholt, B.R., Werner, E.E., 1995. Interaction between food availability and predation mortality mediated by adaptive behavior. Ecology 76 (7), 2230–2234.

Bargmann, G.G., 1977. Instances of copper rockfish consuming a spiny dogfish. Calif. Fish Game 63, 192.

Bernal, D., Lowe, C., 2015. Field studies of elasmobranch physiology. In: Physiology of Elasmobranch Fishes: Structure and Interaction with Environment, vol. 34. p. 311.

Bizzarro, J.J., Yoklavich, M.M., Wakefield, W.W., 2017. Diet composition and foraging ecology of U.S. Pacific Coast groundfishes with applications for fisheries management. Environ. Biol. Fishes 100, 375–393.

Block, B.A., Jonsen, I.D., Jorgensen, S.J., Winship, A.J., Shaffer, S.A., Bograd, S.J., Hazen, E.L., Foley, D.G., Breed, G.A., Harrison, A.L., Ganong, J.E., Swithenbank, A., Castleton, M., Dewar, H., Mate, B.R., Shillinger, G.L., Schaefer, K.M., Benson, S.R., Weise, M.J., Henry, R.W., Costa, D.P., 2011. Tracking apex marine predator movements in a dynamic ocean. Nature 475, 86–90.

Branch, T.A., Watson, R., Fulton, E.A., Jennings, S., McGilliard, C.R., Pablico, G.T., Ricard, D., Tracey, S.R., 2010. The trophic fingerprint of marine fisheries. Nature 468, 431–435.

Bromaghin, J.F., Lance, M.M., Elliot, E.W., Jeffries, S.J., Acevedo-Gutiérrez, A., Kennish, J.M., 2013. New insights into the diets of harbor seals (*Phoca vitulina*) in the Salish Sea revealed by analysis of fatty acid signatures. Fish. Bull. 111, 13–26.

Burgess, G.H., Bruce, B.D., Cailliet, G.M., Goldman, K.J., Grubbs, R.D., Lowe, C.G., MacNeil, M.A., Mollet, H.F., Weng, K.C., O'Sullivan, J.B., 2014. A re-revaluation of the size of the white shark (*Carcharodon carcharias*) population off California, USA. PLoS One 9, e98078.

Campos, B.R., Maxfield, A.F., Jones, G., Riley, R.W., Allen, P.J., Klimley, P.A., Cech Jr., J.J., Kelly, J.T., 2009. Movements of brown smoothhounds, *Mustelus henlei*, in Tomales Bay, California. Environ. Biol. Fishes 85, 3–13.

Carlisle, A.B., Starr, R.M., 2009. Habitat use, residency, and seasonal distribution of female leopard sharks *Triakis semifasciata* in Elkhorn Slough, California. Mar. Ecol. Prog. Ser. 380, 213–228.

Carlisle, A.B., Starr, R.M., 2010. Tidal movements of female leopard sharks (*Triakis semifasciata*) in Elkhorn Slough, California. Environ. Biol. Fishes 89, 31–45.

Carlisle, A.B., Perle, C.R., Goldman, K.J., Block, B.A., 2011. Seasonal changes in depth distribution of salmon sharks (*Lamna ditropis*) in Alaskan waters: implications for foraging ecology. Can. J. Fish. Aquat. Sci. 68, 1905–1921.

Carlisle, A.B., Kim, S.L., Semmens, B.X., Madigan, D.J., Jorgensen, S.J., Perle, C.R., Anderson, S.D., Chapple, T.K., Kanive, P.E., Block, B.A., 2012. Using stable isotope analysis to understand the migration and trophic ecology of northeastern Pacific white sharks (*Carcharodon carcharias*). PLoS One 7. e30492.

Carlisle, A.B., Litvin, S.Y., Hazen, E.L., Madigan, D.J., Goldman, K.J., Lea, R.N., Block, B.A., 2015. Reconstructing habitat use by juvenile salmon sharks links upwelling to strandings in the California Current. Mar. Ecol. Prog. Ser. 525, 217–228.

Cartamil, D.P., Wegner, N.C., Aalbers, S., Sepulveda, C.A., Baquero, A., Graham, J.B., 2010. Diel movement patterns and habitat preferences of the common thresher shark (*Alopias vulpinus*) in the Southern California Bight. Mar. Freshw. Res. 61, 596–604.

Chapple, T., Chambert, T., Kanive, P., Jorgensen, S., Rotella, J., Anderson, S., Carlisle, A., Block, B., 2016. A novel application of multi-event modeling to estimate class segregation in a highly migratory oceanic vertebrate. Ecology 97, 3494–3502.

Christensen, V., Walters, C.J., 2004. Ecopath with Ecosim: methods, capabilities and limitations. Ecol. Model. 172 (2–4), 109–139.

CHAPTER FIVE

Stable Isotope Applications for Understanding Shark Ecology in the Northeast Pacific Ocean

Jonathan C.P. Reum*,[1], Gregory D. Williams[†], Chris J. Harvey[‡]
*University of Washington, School of Aquatic and Fishery Sciences, Seattle, WA, United States
†Pacific States Marine Fisheries Commission, Seattle, WA, United States
‡National Oceanic and Atmospheric Administration, Northwest Fisheries Science Center, Seattle, WA, United States
[1]Corresponding author: e-mail address: jonathan.reum@noaa.gov

A juvenile North Pacific Spiny Dogfish (*Squalus suckleyi*) captured in Puget Sound, Washington. Photo courtesy of Dayv Lowry.

Contents

Advances in Marine Biology, Volume 77
ISSN 0065-2881
http://dx.doi.org/10.1016/bs.amb.2017.06.003

Abstract

Stable isotopes are used to address a wide range of ecological questions and can help researchers and managers better understand the movement and trophic ecology of sharks. Here, we review how shark studies from the Northeast Pacific Ocean (NEP) have employed stable isotopes to estimate trophic level and diet composition and infer movement and habitat-use patterns. To date, the number of NEP shark studies that have used stable isotopes is limited, suggesting that the approach is underutilized. To aid shark researchers in understanding the strengths and limitations of the approach, we provide a brief overview of carbon and nitrogen stable isotope trophic discrimination properties (e.g., change in δ^{15}N between predator and prey), tissue sample preparation methods specific to elasmobranchs, and methodological considerations for the estimation of trophic level and diet composition. We suggest that stable isotopes are a potentially powerful tool for addressing basic questions about shark ecology and are perhaps most valuable when combined and analysed with other data types (e.g., stomach contents, tagging data, or other intrinsic biogeochemical markers).

1. INTRODUCTION

Overfishing, pollution, climate change, and habitat loss are major threats facing sharks throughout the world's oceans, and shark population declines have been observed in several marine ecosystems (Ferretti et al., 2010; Shepherd and Myers, 2005; Ward-Paige et al., 2010). While population trends are available for some species, the status of many others is more difficult to assess because of data limitations ("An Introduction to Modelling Abundance and Demographic Parameters in Shark Populations" by Kacev et al., this volume). However, life history theory suggests that many species of sharks, particularly those with relatively late maturation ages, low fecundity, and low growth rates, are vulnerable to small changes in mortality resulting from anthropogenic factors (Dulvy et al., 2008; Stevens, 2000). Given the vulnerability of sharks, there is a pressing need to better understand their ecology and behaviour in order to develop effective management and conservation strategies (Simpfendorfer et al., 2011).

In the last three decades, stable isotope analysis has emerged as a useful tool in ecological research and has seen perhaps widest application in regard to questions related to the trophic ecology, habitat use, and movement patterns of individuals and populations (Hobson, 1999; Layman et al., 2012; Ramos and González-Solís, 2012). Stable isotope analysis involves using mass spectroscopy to determine the constituent forms of carbon, nitrogen, and other elements that make up the tissues of producers and consumers. The heavy-to-light stable isotope ratios in an organism's tissues directly and predictably reflect its recent ecology, which makes stable isotope analysis a powerful tool (Peterson and Fry, 1987). In the context of shark research, stable isotope studies have yielded insight into a broad range of issues ranging from seasonal foraging and migration patterns to community functional diversity, intrapopulation resource partitioning, and trophic position (e.g., Abrantes and Barnett, 2011; Borrell et al., 2011; Hussey et al., 2014; MacNeil et al., 2005; Matich et al., 2010). The application of stable isotopes to shark research questions in the Northeast Pacific Ocean (NEP), however, is more limited.

In this chapter, we review how stable isotope ratios have been used to study NEP shark populations and highlight research areas that may benefit from their application. Our overarching goal is to draw further attention to how stable isotope analysis can advance the study of sharks in the NEP. To that end, we outline the advantages and limitations of stable isotopes as well as recent methodological advancements that should be considered in the development of future isotope studies. Similar to others, we argue that stable isotopes are perhaps most powerful when used in studies that employ a multidisciplinary framework (Christiansen et al., 2015). In some instances, more than one hypothesis may pose as a plausible explanation for results obtained from stable isotope studies of wild populations. Combining stable isotope results with other data (e.g., telemetry, trace metal, or stomach content data) can provide more powerful evidence to distinguish among alternative hypotheses (Christiansen et al., 2015). We limit our review to applications of carbon and nitrogen stable isotope ratios because they are by far the most widely used in ecological research. However, we refer the interested reader to other resources on the properties and ecological applications of hydrogen, oxygen, and sulphur stable isotopes (e.g., Fry, 2006; Peterson and Fry, 1987).

To set the stage, we first provide a brief overview of nitrogen and carbon stable isotope trophic discrimination properties, that is, the change in δ^{15}N or δ^{13}C between predator and prey. We then focus on two major areas of shark research that draw upon stable isotopes: (1) estimation of trophic level and diet; and (2) habitat association and migration patterns. When possible and relevant, we use examples from research from the NEP.

2. NITROGEN AND CARBON STABLE ISOTOPE TROPHIC DISCRIMINATION

The nitrogen and carbon stable isotope ratios in the tissues of biota are useful ecological tools, in part due to their different trophic discrimination properties. Specifically, in animal predators nitrogen from their diet is assimilated into tissues, but the lighter ^{14}N isotope is preferentially excreted over the heavier ^{15}N. In general, muscle tissues from predators are enriched in $\delta^{15}N$ by ~1–6‰ relative to their prey, using the widely recognized δ-notation for the standardized ratio of the heavy-to-light isotope (see Section 3.1). This strong isotopic discrimination between predator and prey makes $\delta^{15}N$ a good indicator of a consumer's trophic level within a food web. Early meta-analyses suggested an average enrichment for $\delta^{15}N$ between predators and prey between 2.7‰ and 3.4‰ (McCutchan et al., 2003; Post, 2002; Vanderklift and Ponsard, 2003).

In contrast, the heavier isotope ^{13}C is enriched over the lighter ^{12}C only slightly in predators relative to their diet, with $\delta^{13}C$ increasing by ~0.5–1‰ (McCutchan et al., 2003; Peterson and Howarth, 1987). Because of the conservative transfer between prey and predators, $\delta^{13}C$ is often used for tracing energy pathways when the $\delta^{13}C$ values of different energy sources are sufficiently distinct (Michner and Kaufman, 2005). For instance, in shelf marine ecosystems, the $\delta^{13}C$ of pelagic primary producers and consumers generally decreases with distance from shore (Kline and Thomas, 1999; McConnaughey and McRoy, 1979) and the $\delta^{13}C$ values of benthic energy sources are depleted relative to pelagic sources (Davenport and Bax, 2002; Kopp et al., 2015). In these instances, simple two-source mixing models can be used to estimate carbon source contributions to predator diets (Fry, 2006; Phillips, 2001).

Combined, $\delta^{15}N$ and $\delta^{13}C$ values provide a two-dimensional representation of the isotopic niche space occupied by predators (Jackson et al., 2011; Layman et al., 2007). If the isotopic values of individuals from different species or populations overlap, it suggests the potential for overlap in resource use according to diet, habitat, or both. The total area occupied and the level of dispersion of individuals in $\delta^{15}N$–$\delta^{13}C$ space can be analysed quantitatively to compare isotopic niche diversity within species and examine the level of overlap (Layman et al., 2007; Quevedo et al., 2009). For example, in a diverse elasmobranch community occupying an eelgrass-dominated estuary, species overlapped substantially in $\delta^{15}N$–$\delta^{13}C$ space (Vaudo and Heithaus, 2011). However, the spread of individuals in $\delta^{15}N$–$\delta^{13}C$ space was also large, suggesting a high degree of individual specialization that might enable persistence of elasmobranch diversity in that system (Vaudo and Heithaus, 2011). Similar approaches have also been used to examine the intrapopulation structure in stable isotope niche space. If individuals shift habitat or increase prey

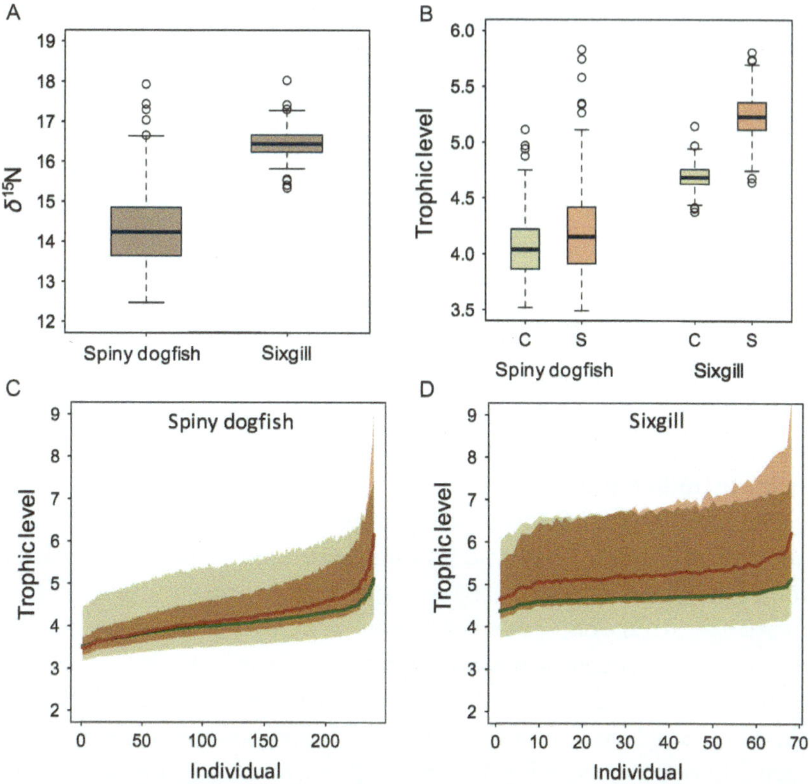

Fig. 1 (A) Box and whisker plot of $\delta^{15}N$ values of white muscle tissue sampled from North Pacific Spiny Dogfish and Bluntnose Sixgill Sharks from Puget Sound, Washington, USA. *Box* corresponds to inner quartile range, *solid horizontal bar* indicates median, and *whiskers* denote data range excluding outliers (indicated by *open circles*). (B) Point estimates of trophic level assuming constant (C, *green*) or scaled (S, *pink*) $\Delta^{15}N$ of dogfish and sixgill individuals. (C and D) Estimated 90% uncertainty percentiles for constant and scaled tropical level estimates for individual dogfish and sixgills, respectively. Individuals are ordered according to their $\delta^{15}N$ values. The median of the simulated trophic level values are indicated by the *solid lines*.

set to 9.0‰ and $\lambda_{baseline}$ to 2.5 (Reum and Essington, 2013). For dogfish with low $\delta^{15}N$ values, the uncertainty around λ_c was twice as great as the uncertainty around λ_s (Fig. 1C), although uncertainty increased at higher $\delta^{15}N$ values for both λ_c and λ_s. For individual sixgills, which generally had higher $\delta^{15}N$ values than dogfish, λ_s uncertainty intervals were generally smaller than those for λ_c (Fig. 1D).

Although there is compelling evidence to adopt scaled $\Delta^{15}N$ for estimating trophic level (Hussey et al., 2014, 2015), the approach is relatively new and consensus on best practices for selecting $\Delta^{15}N$ values has yet to emerge. In this example, we focus on uncertainty related to $\Delta^{15}N$, but we also note that other sources of error including variability in the baseline $\delta^{15}N$ (e.g., Jennings and Van Der Molen, 2015), seasonal variation in prey $\delta^{15}N$, or variation in $\Delta^{15}N$ related to prey nutritional quality (e.g., Sweeting et al., 2006) may impart additional uncertainty into trophic level estimates.

3.2 Diet Composition

In recent years, stable isotope mixing models have become important tools for estimating the composition of predator diets, and Bayesian models in particular have grown in favour because they can account for various error sources, including error in source (prey) stable isotope values and trophic discrimination factors (Moore and Semmens, 2008; Parnell et al., 2010). In addition, versions of Bayesian mixing models allow different trophic discrimination factors between predators and different prey sources, hierarchical variance structures, covariates (e.g., sex, body size, age, location), account for different elemental concentrations among sources, and inclusion of additional information in the form of priors (Parnell et al., 2013; Semmens et al., 2009; Stock and Semmens, 2016). A review on stable isotope mixing models and best practices is provided by Phillips et al. (2014), but we highlight some of the more prominent challenges in the application of stable isotope mixing models to shark diet estimation.

First, the ability of mixing models to provide information on diet composition depends, in part, on whether the sources are sufficiently distinct in terms of their stable isotope values. Because of the high degree of omnivory in marine food webs, the distinction between individual species is often weak, and prey taxa are usually lumped according to functional group or by habitat (e.g., benthic vs pelagic, inshore vs offshore, estuarine vs coastal). Consequently, the resulting estimates of diet proportions usually offer a coarser understanding of prey type relative to standard stomach content analyses. The value of such coarse prey distinctions will therefore depend on the question being posed.

Second, diet estimation is often premised on the assumption that the predator is in isotopic equilibrium with its prey. For shark species that maintain residence in a region for extended periods, it may be feasible to sample prey at matching spatial scales. However, large-scale movement patterns for shark species are also common (Carlisle et al., 2012, 2015a,b). If additional information on movement patterns (e.g., satellite tagging or telemetry data) and lags associated with tissue isotope turnover is available, biases in diet estimation arising from disequilibrium between predator and prey stable isotope concentrations can potentially be accounted for (Hertz et al., 2016), but such data are difficult to obtain and often unavailable.

Third, like estimation of trophic level, the estimated diet proportions derived from mixing models are highly sensitive to the trophic discrimination factor used, indicating a need for careful consideration (Bond and Diamond, 2011). Similar to estimates of trophic level, most published stable isotope-based diet estimates have assumed constant Δ^{15}N. As illustrated earlier, adopting scaled Δ^{15}N values might result in dramatically different conclusions. To illustrate the potential magnitude of this effect, we estimated diet composition using the same stable isotope values analysed in our subadult sixgill trophic level example from Fig. 1. Five prey categories were included as potential source contributions: mobile benthic invertebrates (Dungeness crab, spot prawns, and Pacific giant octopus); pelagic nekton (salmon, North Pacific Spiny Dogfish, Pacific Herring, squid); small demersal fishes (flatfishes, skates, Spotted Ratfish, and medium rockfish, cottids, and gadids 20–50 cm TL); large benthic piscivorous fish (lingcod, large rockfish, cottids, and gadids >50 cm TL); and Harbour seals (Fig. 2). The diet proportions were estimated with a Bayesian stable isotope mixing model

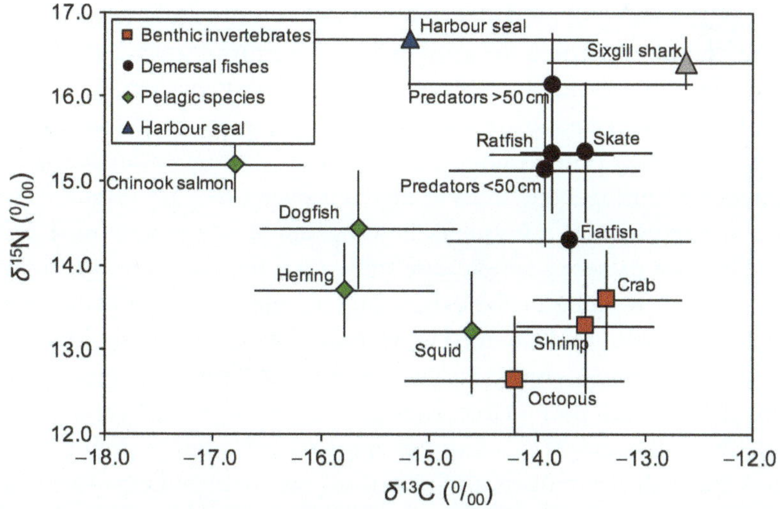

Fig. 2 Mean δ^{15}N and δ^{13}C values of Bluntnose Sixgill Shark (based on white muscle tissue) and their potential prey from Puget Sound, Washington, USA. *Error bars* correspond to standard deviations.

(SIAR, stable isotope analysis in *R*; Parnell et al., 2010) using either a constant Δ^{15}N of 3.4‰ ± 0.98 (Post, 2002) or scaled Δ^{15}N values on the formulas of Hussey et al. (2014) described in the preceding section. Standard deviations for scaled Δ^{15}N values were also calculated following the same simulation procedure used in the trophic level example.

Based on scaled Δ^{15}N, the mean contribution of mobile benthic invertebrates was ~45% of the diet and small demersal fishes, large piscivorous fishes, seals, and pelagic nekton composed ~30%, 10%, 5%, and 2% of diet, respectively (Fig. 3). In contrast, diet estimates based on a constant Δ^{15}N suggest substantially higher dietary reliance on benthic macroinvertebrates (~85%), and less on benthic fishes, pelagic prey, and seals (Fig. 3). Currently, stomach content data for subadult Puget Sound Sixgill Sharks are limited, which makes efforts to judge the appropriateness of assuming constant over scaled Δ^{15}N problematic. However, as mentioned earlier, evidence suggests scaled Δ^{15}N may be more appropriate than the standard assumption of constant Δ^{15}N (Hussey et al., 2014, 2015).

More mechanistic approaches for estimating diet composition are also available but have been underutilized. For example, there are methods of assessing diets and isotopic signatures that do not assume the consumer is in isotopic equilibrium with its diet. Rather, these methods account for the time lags in isotopic equilibration following ontogenetic shifts in diet or other changes in environmental conditions that affect growth (Harvey et al., 2002; Hertz et al., 2016). Long-lived, slow-growing species like sharks require considerable time for growth and tissue turnover in order for their bulk tissues to come into isotopic equilibrium with their diets. This time lag is further complicated by ecological changes, such as if the consumer switches between isotopically distinct habitats or prey communities, or if it experiences ontogenetic shifts in diet as it overcomes gape limits or other functional mechanical constraints limiting the ability to consume certain prey. Here we present a simple case study for a hypothetical North Pacific Spiny Dogfish residing in the NEP using an energy budget-stable isotope modelling framework developed by Harvey et al. (2002), with dogfish bioenergetics parameters from Harvey (2009), diet estimates from Jones and Geen (1977), and prey stable isotope data from a number of sources (Table 2). In this model, a consumer approaches isotopic equilibrium with its prey through accumulation of new tissue (growth) and turnover of existing tissue (maintenance) (Harvey et al., 2002). The model simulates the growth and stable isotope history of the hypothetical dogfish from birth

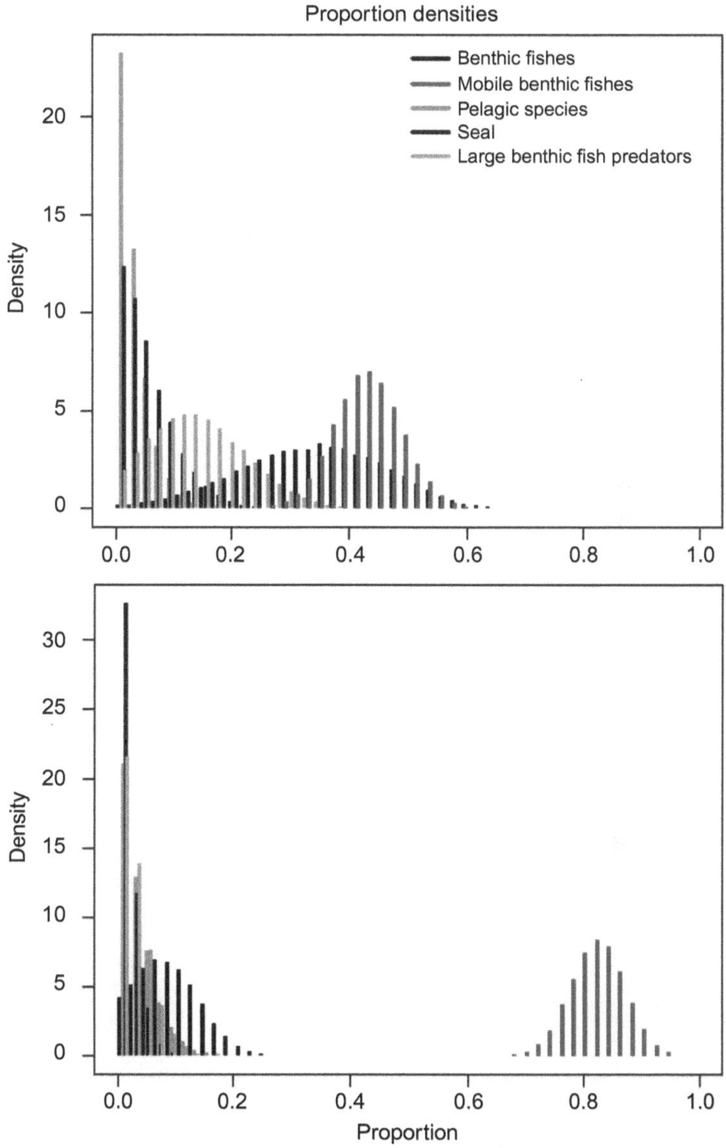

Fig. 3 Diet proportion predictions for subadult Bluntnose Sixgill Shark from Puget Sound, Washington, USA, based on estimates from Bayesian mixing models using scaled (*top panel*) and constant (*bottom*) $\Delta^{15}N$ assumptions. *Adapted from Williams, G.D., Harvey, C.J., Reum, J.C.P., Andrews, K.S., Levin, P.S., in review. Trophic ecology and movement patterns of sixgill sharks in Puget Sound: a stable isotope analysis. Mar. Ecol. Prog. Ser.*

Table 2 Diet Composition and Predator/Prey Stable Isotope Ratios for a Hypothetical North Pacific Spiny Dogfish Residing in the Northeast Pacific Ocean in the Bioenergetics-Stable Isotope Model Simulation

	Mean Stable Isotope Ratios		Diet Proportion by Predator Length Class				
Species Group	$\delta^{15}N$ (‰)	$\delta^{13}C$ (‰)	**<35 cm**	**35–46 cm**	**46–78 cm**	**78–93 cm**	**>93 cm**
Gelatinous zooplankton	11.57	−16.81	0.03	0.03	0.02	0.05	0.03
Crustacean zooplankton	9.36	−17.76	0.15	0.10	0.13	0.13	0.04
Cephalopods	12.80	−16.31	0.05	0.05	0.03	0.06	0.05
Polychaetes	10.50	−19.50	0.05	0.05	0.02	0.01	0.01
Amphipods	10.50	−18.00	0.20	0.10	0.01	0.00	0.00
Brachyurans	11.00	−19.50	0.00	0.00	0.02	0.07	0.11
Euphausiids	9.75	−17.19	0.48	0.46	0.20	0.14	0.06
Pandalid shrimp	14.00	−18.00	0.02	0.05	0.11	0.07	0.05
Other invertebrates	11.24	−17.79	0.00	0.01	0.04	0.03	0.03
Clupeids	12.77	−17.30	0.01	0.16	0.31	0.22	0.26
Other fishes	13.07	−17.89	0.01	0.00	0.10	0.23	0.35
North Pacific Spiny Dogfish (initial)	12.60	−18.01	—	—	—	—	—

North Pacific Spiny Dogfish diets were derived from Jones and Geen (1977), while stable isotope ratios for dogfish and prey items in the Northeast Pacific were derived from Bosley et al. (2004, 2014) and Miller et al. (2008, 2010).

to a maximum age of 75 years, including several major length-based ontogenetic diet shifts.

Fig. 4 shows the model output for both $\delta^{15}N$ and $\delta^{13}C$, and two findings are readily apparent. First, the model clearly demonstrates the importance of assuming constant vs scaled $\Delta^{15}N$ over the full trophic ontogeny of the hypothetical shark (Fig. 4A). Second, the model suggests that the dogfish is often out of equilibrium with its diet, particularly for $\delta^{15}N$ and regardless of whether the constant or scaled $\Delta^{15}N$ is assumed. Only when the shark reaches ~105 cm in length (~30 years of age) does it remain consistently

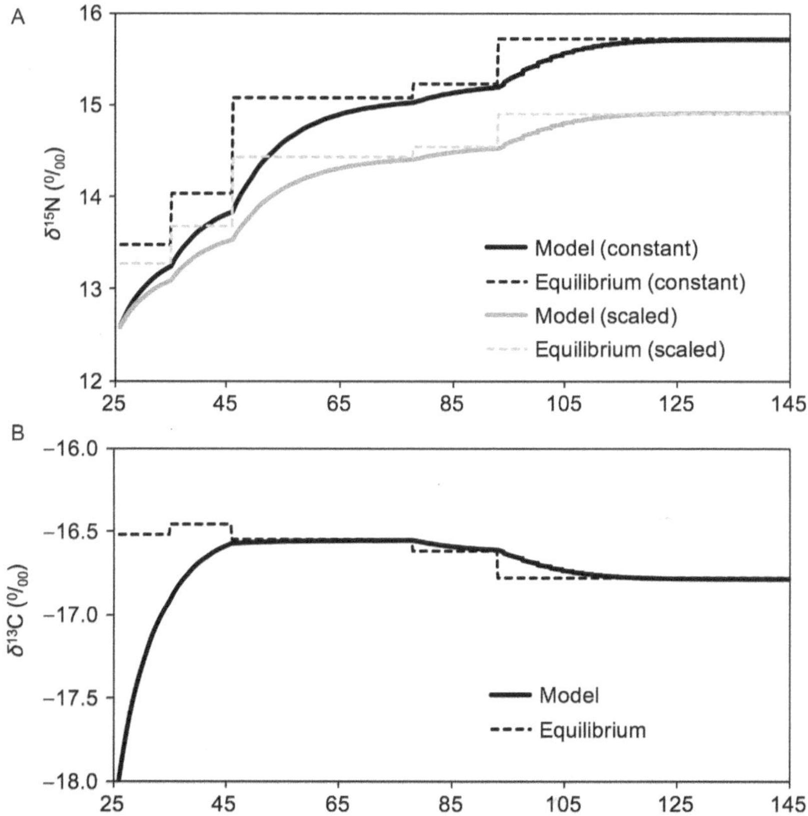

Fig. 4 Simulated ontogeny of stable isotope ratios for a hypothetical North Pacific Spiny Dogfish residing in the NEP, as derived from the bioenergetics-stable isotope model. (A) δ^{15}N values estimated using Δ^{15}N that is either constant (3.4‰; *black*) or scaled (based on Eq. (3); *grey*). (B) δ^{13}C ratios (Δ^{13}C = 1.0‰). *Solid lines* are model outputs, while *dashed lines* are the expected stable isotope ratios if the dogfish were in isotopic equilibrium with its diet. Shifts in the *dashed lines* indicate length-dependent ontogenetic diet shifts (see Table 2).

in isotopic equilibrium with its diet. This suggests that stable isotope-based estimates of diet composition may be biased if the state of isotopic equilibrium is not accounted for (Harvey et al., 2002; Hertz et al., 2016). Although δ^{15}N was the major source of bias in this example, δ^{13}C could be a source of bias in other cases, such as a shark that experiences an ontogenetic transition from a ^{13}C-enriched estuarine habitat to a ^{13}C-depleted offshore habitat. Also, this method can be readily adapted to include estimates of error (Harvey et al., 2002). More generally, the bioenergetics framework could be used for informing aspects of study design. For instance, the potential level of isotopic disequilibrium between predators and prey could be examined given predator growth rates and assumed seasonal or interannual shifts in the isotopic concentrations of prey or diet composition (Harvey et al., 2002).

4. MOVEMENT AND HABITAT USAGE

In the marine environment, differences in the isotopic composition of species near the base of the food web are commonly observed along inshore–offshore, ocean–estuarine–river, and benthic–pelagic habitat continuums (Kline and Thomas, 1999; Lesage et al., 2001; McConnaughey and McRoy, 1979; Miller et al., 2008). Consequently, stable isotope analysis can potentially aid assessment of shark movement or habitat-use patterns if they forage across areas and habitats that differ sufficiently in terms of prey isotopic composition. For instance, Carlisle et al. (2015a) obtained stable isotope estimates for the potential prey of Salmon Shark from oceanic and neritic habitats in the California Current ecosystem. They found that prey clustered according to three habitat types (inshore shelf, shelf slope, and oceanic mesopelagic) and were able to use a Bayesian mixing model to estimate the contribution of prey from each habitat type in tissues that integrated diet over successively longer periods of time (liver, vertebrate, and muscle). Overall, a shift from oceanic mesopelagic prey to inshore shelf prey was observed (Carlisle et al., 2015a). Similarly, the relative use of three large ecoregions in the NEP by White Sharks was inferred using a stable isotope mixing model (Carlisle et al., 2012). In that study, White Sharks were found to forage more intensively in coastal waters relative to offshore, pelagic habitats.

In other NEP shark studies, which lacked spatial isotope information on different potential prey communities, stable isotope concentrations in tissues with different turnover times suggested temporal shifts in habitat, although not conclusively. In populations of North Pacific Spiny Dogfish, and small-bodied Shortfin Mako and White Sharks, differences in stable isotope ratios between tissues with relatively slow (cartilage or muscle) and fast (liver or blood) turnover rates suggested that individuals may have only recently arrived (within a few weeks to months) to the region in which they were captured for sampling (Malpica-Cruz et al., 2013; Reum and Essington, 2013). Conversely, similarity in isotope concentrations in tissues with different turnover rates in large-bodied Shortfin Mako and White Sharks suggested long-term residency (Malpica-Cruz et al., 2013). While such studies offer important exploratory information and a basis for developing more targeted future studies, the absence of spatial information on prey stable isotope values limits the potential for inference with regard to movement. Rather, seasonal shifts in diet may also be plausible explanations for the observed patterns (Malpica-Cruz et al., 2013; Reum and Essington, 2013). The possibility of multiple plausible hypotheses to explain patterns in stable isotope data is common, particularly for large, highly mobile predators and should be clearly articulated in stable isotope studies (Christiansen et al., 2015).

The application of stable isotopes to questions regarding movement and habitat use is perhaps most powerful when combined with other data types, such as information obtained from satellite tagging, acoustic telemetry, or contaminant analyses. For example, Carlisle et al. (2015b) combined stable isotope data with a large, multiyear satellite tagging data set to examine ontogenetic shifts in habitat use by Salmon Shark. In that study, $\delta^{15}N$ and $\delta^{13}C$ chronologies were obtained for individual sharks by sampling vertebral annuli, and the satellite tagging information permitted estimation of the approximate time spent in four large ecoregions in the NEP delineated as northern and southern neritic and oceanic areas. The contribution of prey from each region was estimated for each age class, assuming noninformative priors for age classes less than year 8. For older age classes, which corresponded to the ages of animals included in the tracking study, the time spent in each region based on the tagging data was incorporated in the mixing model as informative priors. For adults, the model suggests higher reliance on prey from the northern neritic ecoregion relative to a model that did not incorporate tagging data.

Additionally, stable isotopes can be paired with other intrinsic biogeochemical markers, including trace elements or contaminants, to infer movement patterns (Ramos and González-Solís, 2012). While examples of shark studies that evaluate movement patterns using both stable isotopes and contaminants are limited, the approach has been employed successfully in other taxa. For instance, in the Northeast Pacific, the winter geographic range of southern resident Killer Whales *Orcinus orca* was inferred using stable isotopes and regionally specific, persistent organic pollutant (POP) concentrations in their primary prey, Pacific salmon *Oncorhynchus* spp. (Krahn et al., 2007; Mongillo et al., 2016). To illustrate the utility of combining stable isotope data with contaminant information, we present a recent analysis of Bluntnose Sixgill Shark tissues collected from individuals from Puget Sound, Washington. The analysis helps clarify long-standing questions regarding residency and migrations patterns of Puget Sound Sixgills.

Adult sixgills occur in deepwater areas over the continental shelf and upper slope (Ebert, 2003) and are rarely observed or captured in shallow, nearshore habitats, where juveniles and subadults are typically documented (Williams et al., 2010). Acoustic tagging studies indicate that subadult sixgill sharks are resident in Puget Sound year-round and forage over relatively small regions given their size, but exhibit larger seasonal movement patterns within Puget Sound (Andrews et al., 2007, 2010). Acoustic tagging evidence suggests sixgill sharks leave Puget Sound upon reaching maturity (Andrews et al., 2010). Overall, subadult sixgill shark muscle tissues collected from individuals in Puget Sound are enriched in $\delta^{13}C$ (Fig. 5), suggesting dependence on nearshore, benthic production. Furthermore, their POP concentrations reflect low ΣDDT/ΣPCB ratios comparable to Chinook Salmon *Oncorhynchus tshawytscha* stocks and a pod of southern resident Killer Whales, which reside and feed year-round in Puget Sound where PCB levels are relatively high (Fig. 6; Mongillo et al., 2016).

In contrast, muscle tissues obtained from an adult female and three prenatal pups that stranded in a shallow inlet of south Puget Sound were relatively depleted in $\delta^{13}C$, suggesting reliance on prey from offshore

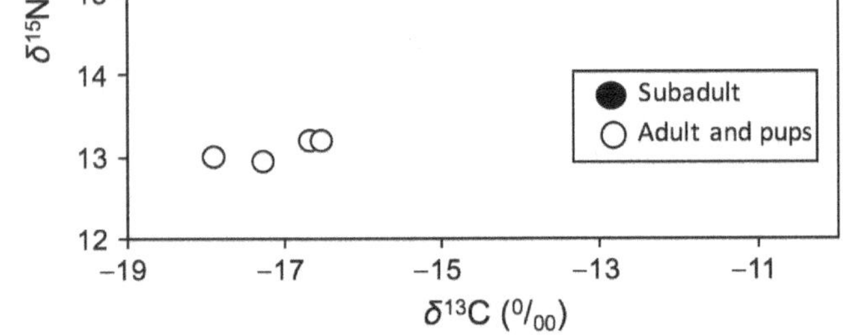

Fig. 5 $\delta^{15}N$–$\delta^{13}C$ biplot of white muscle tissue samples obtained from subadult Bluntnose Sixgill Shark (*filled circles; n*=43) and one adult female with three pups (*open circles*) in Puget Sound, Washington, USA. *Adapted from Williams, G.D., Harvey, C.J., Reum, J.C.P., Andrews, K.S., Levin, P.S., in review. Trophic ecology and movement patterns of sixgill sharks in Puget Sound: a stable isotope analysis. Mar. Ecol. Prog. Ser.*

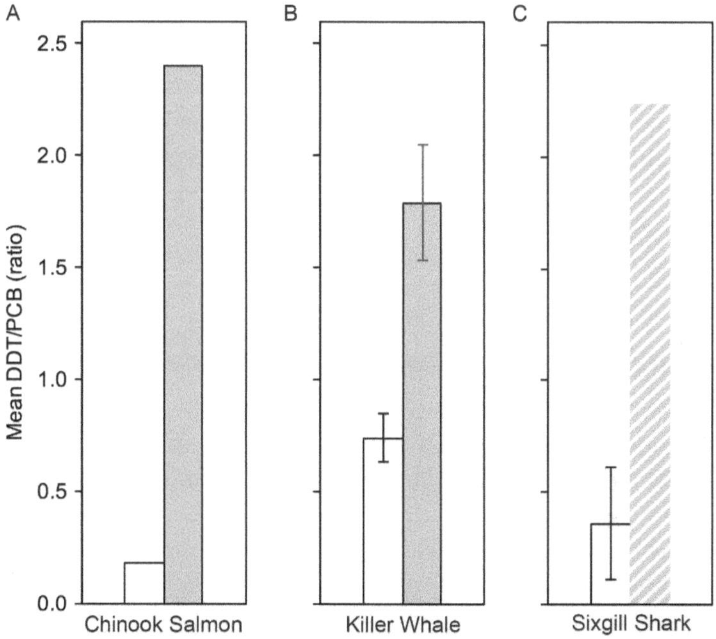

Fig. 6 Persistent organic pollutant (POP) concentration ratios (mean ΣDDTs/ΣPCBs ratio ± SE) of: (A) Chinook Salmon from Puget Sound, Washington (*white bar*, $n=34$) and California (*grey*, $n=29$) coastal waters (Sacramento and San Joaquin estuaries; Krahn et al., 2007; Mongillo et al., 2016); (B) southern resident Killer Whales that primarily forage in Washington and southern British Columbia inland marine and open coast waters year-round (*white bar*, $n=5$) and migratory Killer Whales that also forage in California coastal waters (*grey bar*, $n=14$; Krahn et al., 2007; unpublished NWFSC data), and (C) Bluntnose Sixgill Shark sampled from Puget Sound. Chinook Salmon are a principle prey of Killer Whales, and their POP ratios reflect those of their prey (Ylitalo et al. NWFSC, unpublished data). For sixgills, acoustic tagging evidence suggests that subadults (*white bar*, $n=12$) reside in Puget Sound year-round until reaching maturity, and their POP ratios are consistent with Chinook Salmon resident in Puget Sound. POP ratios obtained from one adult female sixgill (stripped *grey bar*) sampled in southern Puget Sound, however, differ substantially from the subadults but are consistent with the levels observed in migratory Killer Whales and their California Chinook Salmon prey, suggesting that it also likely migrated from coastal shelf waters off California. *Error bars* correspond to standard error.

continental shelf or slope waters (Fig. 5). The adult female's POP concentrations were also higher than the subadult sixgill, reflecting increased consumption of prey containing higher relative DDTs (Fig. 6). Previous contaminant studies on migratory west coast Killer Whales indicate that similar high DDT concentrations reflect feeding on Chinook Salmon from California waters (Fig. 6; Mongillo et al., 2016; O'Neill et al., 2006). Taken together, these findings provide compelling evidence that Puget Sound is an important nursery for Bluntnose Sixgill Shark, and that some adult females may migrate from continental shelf waters off California to the inner basins of Puget Sound for parturition. Understanding the migratory patterns of sharks remains a major challenge to their successful management and conservation, and stable isotopes coupled with other tracers can be potentially informative tools.

5. FUTURE RESEARCH DIRECTIONS

Much of this chapter has focused on traditional applications of stable isotope analysis in ecological studies, in particular the use of $\delta^{13}C$ and $\delta^{15}N$ values of tissues (white muscle, liver, etc.) to address questions regarding trophic ecology and movement. However, other techniques that expand the power of conventional stable isotopes analyses are rapidly emerging, and we briefly summarize some of those methods here, providing shark-related examples when possible.

5.1 Complementary Tracer Methods

As illustrated in the earlier example of contaminant chemistry in Bluntnose Sixgill Shark, stable isotope analysis can be paired with other kinds of intrinsic tracers to increase the resolution of potential sources of production. One prominent example is analysis of naturally occurring fatty acids (FAs). FAs are a highly diverse group of chemicals, many of them transfer in predictable manners from prey to predator, and some can be traced back to particular primary producers (e.g., Budge et al., 2006; Iverson, 2009; Kelly and Scheibling, 2012). These traits can make FAs effective tracers of food web pathways and indicators of trophic guilds, particularly when profiles of multiple FAs are examined from different species in a community. FAs can complement information from bulk tissue SIA in cases where, for example, FA profiles discriminate more finely between different production sources than $\delta^{13}C$, but $\delta^{15}N$ provides a more robust indicator of trophic level than FAs (Kelly and Scheibling, 2012). This was demonstrated by Wai et al. (2012), who analysed bulk tissue $\delta^{13}C$, $\delta^{15}N$, and FAs to quantify size-, species-, and habitat-based diet differences and contributions of terrestrial, pelagic, and benthic carbon to two shark species in the Pearl River Estuary of southern China.

Still greater power is added by analysing stable isotope ratios of specific compounds within consumer tissues (Boecklen et al., 2011). Compound-specific stable isotope analysis has been applied to FAs (e.g., Kelly and Scheibling, 2012; Wai et al., 2012) and amino acids (AAs; e.g., McMahon and McCarthy, 2016). The added information derived from stable isotope analysis of numerous individual FAs or AAs may provide far greater source discrimination and pose fewer complicating assumptions than stable isotope analysis of composite bulk tissues (Boecklen et al., 2011). Compound-specific stable isotope analysis is beginning to be applied to shark research. For example, Hoen et al. (2014) examined $\delta^{15}N$ trophic discrimination of several AAs in long-term feeding experiments on three shark species. In their study of sharks in the Pearl River Estuary, Wai et al. (2012) analysed $\delta^{13}C$ in FAs to further resolve the relative contributions of different carbon sources in shark diets.

While these complementary methods offer great promise to enhance our understanding of shark feeding and habitat use, much foundational work must be done before they can be utilized fully. In particular, controlled feeding experiments to better understand the dynamics of individual FAs, AAs, and other compounds under different conditions are needed to properly interpret results from the field (Iverson, 2009; Ramos and González-Solís, 2012). This fact does not preclude use of complementary tracer studies right now, but it highlights the need for care in interpreting results and understanding the limitations, caveats, and best practices associated with employment of these methods (Boecklen et al., 2011; Iverson, 2009; Kelly and Scheibling, 2012).

5.2 Other Approaches

Still other multidisciplinary approaches that incorporate stable isotope analyses have improved understanding of fish ecology in recent years and could also be applied to research on NEP sharks. For example, identifying essential fish habitat (EFH) is a critical aspect of fish conservation and management, and EFH assessment includes understanding key mechanistic relationships between that habitat and the life history stages of the fishes that depend on it. This can be difficult for highly mobile species (Deegan et al., 2000), which sharks very often are. Quantifying traits of a fish population—survival rates, reproduction, growth—while in different habitats is critical to understanding the value of different habitats to overall population condition and growth rate, and stable isotope analyses may help to provide linkages to different habitats during different windows in a fish's life cycle. In coastal habitats along the Atlantic coast of the United States, Weinstein et al. (2009) and Litvin et al. (2014) used stable isotope and FA analysis to discern habitat quality in different life history stages of several teleosts across habitat disturbance gradients. Studies of this sort could help researchers to prioritize habitats for protection or restoration in support of shark populations.

Ecosystem models are another tool that has seen increasing use in marine resource management and conservation. Such models vary widely in complexity, but very often simulate the internal dynamics of a multispecies food web that is further influenced by external forcing, from climate drivers, nutrient cycling, and/or anthropogenic activities such as fishing, habitat modification, or pollution. Many such ecosystem models exist for food webs in the NEP and include sharks as predators (e.g., Harvey et al., 2012; Kaplan et al., 2012; Ruzicka et al., 2016; Samhouri et al., 2009). The dynamics of species in these models are highly dependent upon the assumptions that modellers make about their trophic interactions, and these assumptions are very often based on limited and caveat-laden stomach content data. Stable isotope analysis, particularly if coupled with other tracers or models to provide robust diet estimates, could greatly improve the information quality for sharks and other species in these food web models and thereby reduce model uncertainty. Improvements in model performance could, in turn, help us to better anticipate how changes in fishing, environmental conditions, food web structure, or other ecosystem-scale factors will affect shark populations in the NEP; similar conclusions have been drawn for sharks in other regions of the world (e.g., Barria et al., 2015).

6. CONCLUSION

The field of stable isotope ecology is relatively new, but has rapidly expanded in the last three decades. The sophistication of quantitative methods for analysing and modelling stable isotope ratios has increased, broadening the potential scope for inference (e.g., Boecklen et al., 2011; Layman et al., 2012; Parnell et al., 2013; Semmens et al., 2009). At the same time, workers are increasingly combining stable isotope analysis with other data sources in multidisciplinary frameworks (e.g., Carlisle et al., 2012, 2015a). Doing so provides more robust evidence for distinguishing between alternative hypotheses. Given the need to better understand the basic ecology of sharks in the NEP, researchers and managers alike could benefit from considering how stable isotope analysis might help answer the pressing questions they face regarding the movement patterns, habitat usage, and trophic roles of sharks.

ACKNOWLEDGEMENTS

The authors thank Eric Ward, Kelly Andrews, and an anonymous reviewer for comments and suggestions on earlier versions of the manuscript and Dan Lomax and Gina Ylitalo (NWFSC) for sharing contaminant data.

REFERENCES

Abrantes, K.G., Barnett, A., 2011. Intrapopulation variations in diet and habitat use in a marine apex predator, the broadnose sevengill shark *Notorynchus cepedianus*. Mar. Ecol. Prog. Ser. 442, 133–148.

Anderson, W.G., Good, J.P., Pillans, R.D., Hazon, N., Franklin, C.E., 2005. Hepatic urea biosynthesis in the euryhaline elasmobranch *Carcharhinus leucas*. J. Exp. Zool. A Comp. Exp. Biol. 303, 917–921.

Andrews, A.G., Foy, R.J., 2009. Geographical variation in the carbon and nitrogen stable isotope ratios in Spiny Dogfish in the Northeastern Pacific Ocean. In: Gallucci, V.F., McFarlane, G.A., Bargmann, G.G. (Eds.), Biology and Management of Dogfish Sharks. American Fisheries Society, Bethesda, MD, pp. 269–276.

Andrews, K.S., Levin, P.S., Katz, S.L., Farrer, D., Gallucci, V.F., Bargmann, G., 2007. Acoustic monitoring of sixgill shark movements in Puget Sound: evidence for localized movement. Can. J. Zool. 85, 1136–1142.

Andrews, K.S., Williams, G.D., Levin, P.S., 2010. Seasonal and ontogenetic changes in movement patterns of sixgill sharks. PLoS One 5, e12549.

Barria, C., Coll, M., Navarro, J., 2015. Unravelling the ecological role and trophic relationships of uncommon and threatened elasmobranchs in the western Mediterranean Sea. Mar. Ecol. Prog. Ser. 539, 225–240.

Baum, J.K., Myers, R.A., Kehler, D.G., Worm, B., Harley, S.J., Doherty, P.A., 2003. Collapse and conservation of shark populations in the Northwest Atlantic. Science 299, 389–392.

Boecklen, W.J., Yarnes, C.T., Cook, B.A., James, A.C., 2011. On the use of stable isotopes in trophic ecology. Annu. Rev. Ecol. Evol. Syst. 42, 411–440.

Bond, A.L., Diamond, A.W., 2011. Recent Bayesian stable-isotope mixing models are highly sensitive to variation in discrimination factors. Ecol. Appl. 21, 1017–1023.

Borrell, A., Aguilar, A., Gazo, M., Kumarran, R., Cardona, L., 2011. Stable isotope profiles in whale shark (*Rhincodon typus*) suggest segregation and dissimilarities in the diet depending on sex and size. Environ. Biol. Fish. 92, 559–567.

Bosley, K.L., Lavelle, J.W., Brodeur, R.D., Wakefield, W.W., Emmett, R.L., Baker, E.T., Rehmke, K.M., 2004. Biological and physical processes in and around Astoria submarine Canyon, Oregon, USA. J. Mar. Sys. 50, 21–37.

Bosley, K.L., Miller, T.W., Brodeur, R.D., Bosley, K.M., Van Gaest, A., Elz, A., 2014. Feeding ecology of juvenile rockfishes off Oregon and Washington based on stomach content and stable isotope analyses. Mar. Biol. 161, 2381–2393.

Budge, S.M., Iverson, S.J., Koopman, H.N., 2006. Studying trophic ecology in marine ecosystems using fatty acids: a primer on analysis and interpretation. Mar. Mamm. Sci. 22, 759–801.

Cabana, G., Rasmussen, J.B., 1994. Modeling food-chain structure and contaminant bioaccumulation using stable nitrogen isotopes. Nature 372, 255–257.

Carlisle, A.B., Kim, S.L., Semmens, B.X., Madigan, D.J., Jorgensen, S.J., Perle, C.R., Anderson, S.D., Chapple, T.K., Kanive, P.E., Block, B.A., 2012. Using stable isotope analysis to understand the migration and trophic ecology of northeastern Pacific white sharks (*Carcharodon carcharias*). PLoS One 7, e30492.

Carlisle, A.B., Goldman, K.J., Litvin, S.Y., Madigan, D.J., Bigman, J.S., Swithenbank, A.M., Kline, T.C., Block, B.A., 2015a. Stable isotope analysis of vertebrae reveals ontogenetic changes in habitat in an endothermic pelagic shark. Proc. R. Soc. B 282, 20141446.

Carlisle, A.B., Litvin, S.Y., Hazen, E.L., Madigan, D.J., Goldman, K.J., Lea, R.N., Block, B.A., 2015b. Reconstructing habitat use by juvenile salmon sharks links upwelling to strandings in the California Current. Mar. Ecol. Prog. Ser. 525, 217–228.

Carlisle, A.B., Litvin, S.Y., Madigan, D.J., Lyons, K., Bigman, J.S., Ibarra, M., Bizzarro, J.J., 2016. Interactive effects of urea and lipid content confound stable isotope analysis in elasmobranch fishes. Can. J. Fish. Aquat. Sci. 74, 419–428.

Caut, S., Angulo, E., Courchamp, F., 2009. Variation in discrimination factors (Δ^{15}N and Δ^{13}C): the effect of diet isotopic values and applications for diet reconstruction. J. Appl. Ecol. 46, 443–453.

Caut, S., Jowers, M.J., Michel, L., Lepoint, G., Fisk, A.T., 2013. Diet-and tissue-specific incorporation of isotopes in the shark *Scyliorhinus stellaris*, a North Sea mesopredator. Mar. Ecol. Prog. Ser. 492, 185–198.

Christiansen, H., Fisk, A., Hussey, N., 2015. Incorporating stable isotopes into a multidisciplinary framework to improve data inference and their conservation and management application. Afr. J. Mar. Sci. 37, 189–197.

Churchill, D.A., Heithaus, M.R., Grubbs, R.D., 2015. Effects of lipid and urea extraction on δ^{15}N values of deep-sea sharks and hagfish: can mathematical correction factors be generated? Deep-Sea Res. II. Top. Stud. Oceanogr. 115, 103–108.

Courtney, D., Foy, R., 2012. Pacific sleeper shark *Somniosus pacificus* trophic ecology in the eastern North Pacific Ocean inferred from nitrogen and carbon stable-isotope ratios and diet. J. Fish Biol. 80, 1508–1545.

Davenport, S.R., Bax, N.J., 2002. A trophic study of a marine ecosystem off southeastern Australia using stable isotopes of carbon and nitrogen. Can. J. Fish. Aquat. Sci. 59, 514–530.

Deegan, L.A., Hughes, J.E., Rountree, R.A., 2000. Salt marsh ecosystem support of marine transient species. In: Weinstein, M.P., Kreeger, D. (Eds.), Concepts and Controversies in Tidal Marsh Ecology. Kluwer Academic, Dordrecht, Netherlands, pp. 333–365.

Deniro, M.J., Epstein, S., 1977. Mechanism of carbon isotope fractionation associated with lipid-synthesis. Science 197, 261–263.

Dennis, C.A., Macneil, M.A., Rosati, J.Y., Pitcher, T.E., Fisk, A.T., 2010. Diet discrimination factors are inversely related to δ^{15}N and δ^{13}C values of food for fish under controlled conditions. Rapid Commun. Mass Spectrom. 24, 3515–3520.

Dulvy, N.K., Baum, J.K., Clarke, S., Compagno, L.J.V., Cortes, E., Domingo, A., Fordham, S., Fowler, S., Francis, M.P., Gibson, C., Martinez, J., Musick, J.A., Soldo, A., Stevens, J.D., Valenti, S., 2008. You can swim but you can't hide: the global status and conservation of oceanic pelagic sharks and rays. Aquat. Conserv. 18, 459–482.

Ebert, D., 2003. Sharks, Rays and Chimaeras of California. University of California Press, Berkeley, California.

Ferretti, F., Worm, B., Britten, G.L., Heithaus, M.R., Lotze, H.K., 2010. Patterns and ecosystem consequences of shark declines in the ocean. Ecol. Lett. 13, 1055–1071.

Fry, B., 2006. Stable Isotope Ecology. Springer, New York, NY.

Gascuel, D., 2005. The trophic-level based model: a theoretical approach of fishing effects on marine ecosystems. Ecol. Model. 189, 315–322.

Hammerschlag-Peyer, C.M., Yeager, L.A., Araújo, M.S., Layman, C.A., 2011. A hypothesis-testing framework for studies investigating ontogenetic niche shifts using stable isotope ratios. PLoS One 6, e27104.

Harvey, C.J., 2009. Effects of temperature change on demersal fishes in the California Current: a bioenergetics approach. Can. J. Fish. Aquat. Sci. 66, 1449–1461.

Harvey, C.J., Hanson, P.C., Essington, T.E., Brown, P.B., Kitchell, J.F., 2002. Using bioenergetics models to predict stable isotope ratios in fishes. Can. J. Fish. Aquat. Sci. 59, 115–124.

Harvey, C.J., Williams, G.D., Levin, P.S., 2012. Food web structure and trophic control in central Puget Sound. Estuar. Coasts 35, 821–838.

Hazon, N., Wells, A., Pillans, R.D., Good, J.P., Anderson, W.G., Franklin, C.E., 2003. Urea based osmoregulation and endocrine control in elasmobranch fish with special reference to euryhalinity. Comp. Biochem. Physiol. B 136, 685–700.

Hertz, E., Trudel, M., El-Sabaawi, R., Tucker, S., Dower, J.F., Beacham, T.D., Edwards, A.M., Mazumder, A., 2016. Hitting the moving target: modelling ontogenetic shifts with stable isotopes reveals the importance of isotopic turnover. J. Anim. Ecol. 85, 681–691.

Hobson, K.A., 1999. Tracing origins and migration of wildlife using stable isotopes: a review. Oecologia 120, 314–326.

Hoen, D.K., Kim, S.L., Hussey, N.E., Wallsgrove, N.J., Drazen, J.C., Popp, B.N., 2014. Amino acid ^{15}N trophic enrichment factors of four large carnivorous fishes. J. Exp. Mar. Biol. Ecol. 453, 76–83.

Hoffman, J.C., Sutton, T.T., 2010. Lipid correction for carbon stable isotope analysis of deep-sea fishes. Deep-Sea Res. Pt. I 57, 956–964.

Hussey, N.E., Olin, J.A., Kinney, M.J., McMeans, B.C., Fisk, A.T., 2012. Lipid extraction effects on stable isotope values ($\delta^{13}C$ and $\delta^{15}N$) of elasmobranch muscle tissue. J. Exp. Mar. Biol. Ecol. 434, 7–15.

Hussey, N.E., MacNeil, M.A., McMeans, B.C., Olin, J.A., Dudley, S.F., Cliff, G., Wintner, S.P., Fennessy, S.T., Fisk, A.T., 2014. Rescaling the trophic structure of marine food webs. Ecol. Lett. 17, 239–250.

Hussey, N.E., MacNeil, M.A., Siple, M.C., Popp, B.N., Dudley, S.F., Fisk, A.T., 2015. Expanded trophic complexity among large sharks. Food Webs 4, 1–7.

Hyslop, E.J., 1980. Stomach contents analysis: a review of methods and their application. J. Fish Biol. 17, 411–429.

Iverson, S.J., 2009. Tracing aquatic food webs using fatty acids: from qualitative indicators to quantitative determination. In: Arts, M.T., Brett, M.T., Kainz, M. (Eds.), Lipids in Aquatic Ecosystems. Springer, New York, pp. 281–307.

Jackson, A.L., Inger, R., Parnell, A.C., Bearhop, S., 2011. Comparing isotopic niche widths among and within communities: SIBER—Stable Isotope Bayesian Ellipses in R. J. Anim. Ecol. 80, 595–602.

Jennings, S., Blanchard, J.L., 2004. Fish abundance with no fishing: predictions based on macroecological theory. J. Anim. Ecol. 73, 632–642.

Jennings, S., Dulvy, N.K., 2005. Reference points and reference directions for size-based indicators of community structure. ICES J. Mar. Sci. 62, 397–404.

Jennings, S., Van Der Molen, J., 2015. Trophic levels of marine consumers from nitrogen stable isotope analysis: estimation and uncertainty. ICES J. Mar. Sci. 72, 2289–2300.

Jones, B.C., Geen, G.H., 1977. Food and feeding of spiny dogfish (*Squalus acanthias*) in British Columbia waters. J. Fish. Res. B Can. 34, 2067–2078.

Kaplan, I.C., Horne, P.J., Levin, P.S., 2012. Screening California Current fishery management scenarios using the Atlantis end-to-end ecosystem model. Prog. Oceanogr. 102, 5–18.

Kelly, J.R., Scheibling, R.E., 2012. Fatty acids as dietary tracers in benthic food webs. Mar. Ecol. Prog. Ser. 446, 1–22.

Kerr, L.A., Andrews, A.H., Cailliet, G.M., Brown, T.A., Coale., K.H., 2006. Investigations of $\Delta^{14}C$, $\delta^{13}C$, and $\delta^{15}N$ in vertebrae of white shark (*Carcharodon carcharias*) from the eastern North Pacific Ocean. Environ. Biol. Fish. 77, 337–353.

Kiljunen, M., Grey, J., Sinisalo, T., Harrod, C., Immonen, H., Jones, R.I., 2006. A revised model for lipid-normalizing $\delta^{13}C$ values from aquatic organisms, with implications for isotope mixing models. J. Appl. Ecol. 43, 1213–1222.

Kim, S.L., Koch, P.L., 2012. Methods to collect, preserve, and prepare elasmobranch tissues for stable isotope analysis. Environ. Biol. Fish. 95, 53–63.

Kim, S.L., Tinker, M.T., Estes, J.A., Koch, P.L., 2012. Ontogenetic and among-individual variation in foraging strategies of northeast Pacific white sharks based on stable isotope analysis. PLoS One 7, e45068.

Kline, J., Thomas, C., 1999. Temporal and spatial variability of $^{13}C/^{12}C$ and $^{15}N/^{14}N$ in pelagic biota of Prince William Sound, Alaska. Can. J. Fish. Aquat. Sci. 56, 94–117.

Kopp, D., Lefebvre, S., Cachera, M., Villanueva, M.C., Ernande, B., 2015. Reorganization of a marine trophic network along an inshore–offshore gradient due to stronger pelagic–benthic coupling in coastal areas. Prog. Oceanogr. 130, 157–171.

Krahn, M.M., Herman, D.P., Matkin, C.O., Durban, J.W., Barrett-Lennard, L., Burrows, D.G., Dahlheim, M.E., Black, N., LeDuc, R.G., Wade, P.R., 2007. Use of chemical tracers in assessing the diet and foraging regions of eastern North Pacific killer whales. Mar. Environ. Res. 63, 91–114.

Layman, C.A., Arrington, D.A., Montana, C.G., Post, D.M., 2007. Can stable isotope ratios provide for community-wide measures of trophic structure? Ecology 88, 42–48.

Layman, C.A., Araujo, M.S., Boucek, R., Hammerschlag-Peyer, C.M., Harrison, E., Jud, Z.R., Matich, P., Rosenblatt, A.E., Vaudo, J.J., Yeager, L.A., 2012. Applying stable isotopes to examine food-web structure: an overview of analytical tools. Biol. Rev. 87, 545–562.

Lesage, V., Hammill, M.O., Kovacs, K.M., 2001. Marine mammals and the community structure of the Estuary and Gulf of St Lawrence, Canada: evidence from stable isotope analysis. Mar. Ecol. Prog. Ser. 210, 203–221.

Li, Y., Zhang, Y., Hussey, N.E., Dai, X., 2016. Urea and lipid extraction treatment effects on $\delta^{15}N$ and $\delta^{13}C$ values in pelagic sharks. Rapid Commun. Mass Spectrom. 30, 1–8.

Litvin, S.Y., Weinstein, M.P., Guida, V.G., 2014. Habitat utilization patterns determine the physiological condition of *Cynoscion regalis* during estuarine residency. Mar. Ecol. Prog. Ser. 510, 87–99.

Logan, J.M., Lutcavage, M.E., 2008. A comparison of carbon and nitrogen stable isotope ratios of fish tissues following lipid extractions with non-polar and traditional chloroform/methanol solvent systems. Rapid Commun. Mass Spectrom. 22, 1081–1086.

Logan, J.M., Lutcavage, M.E., 2010. Stable isotope dynamics in elasmobranch fishes. Hydrobiologia 644, 231–244.

Logan, J.M., Jardine, T.D., Miller, T.J., Bunn, S.E., Cunjak, R.A., Lutcavage, M.E., 2008. Lipid corrections in carbon and nitrogen stable isotope analyses: comparison of chemical extraction and modelling methods. J. Anim. Ecol. 77, 838–846.

Lyons, K., Preti, A., Madigan, D.J., Wells, R.J.D., Blasius, M.E., Snodgrass, O.E., Kacev, D., Harris, J.D., Dewar, H., Kohin, S., MacKenzie, K., Lowe, C.G., 2015. Insights into the life history and ecology of a large shortfin mako shark *Isurus oxyrinchus* captured in southern California. J. Fish Biol. 87, 200–211.

Mackintosh, C.E., Maldonado, J., Hongwu, J., Hoover, N., Chong, A., Ikonomou, M.G., Gobas, F.A.P.C., 2004. Distribution of phthalate esters in a marine aquatic food web: comparison to polychlorinated biphenyls. Environ. Sci. Technol. 38, 2011–2020.

MacNeil, M.A., Skomal, G.B., Fisk, A.T., 2005. Stable isotopes from multiple tissues reveal diet switching in sharks. Mar. Ecol. Prog. Ser. 302, 199–206.

Madigan, D.J., Carlisle, A.B., Dewar, H., Snodgrass, O.E., Litvin, S.Y., Micheli, F., Block, B.A., 2012. Stable isotope analysis challenges wasp-waist food web assumptions in an upwelling pelagic ecosystem. Sci. Rep. 2, 654.

Malpica-Cruz, L., Herzka, S.Z., Sosa-Nishizaki, O., Escobedo-Olvera, M.A., 2013. Tissue-specific stable isotope ratios of shortfin mako (*Isurus oxyrinchus*) and white (*Carcharodon carcharias*) sharks as indicators of size-based differences in foraging habitat and trophic level. Fish. Oceanogr. 22, 429–445.

Mancinelli, G., Vizzini, S., Mazzola, A., Maci, S., Basset, A., 2013. Cross-validation of $\delta^{15}N$ and FishBase estimates of fish trophic position in a Mediterranean lagoon: the importance of the isotopic baseline. Estuar. Coast. Shelf Sci. 135, 77–85.

Matich, P., Heithaus, M.R., Layman, C.A., 2010. Contrasting patterns of individual specialization and trophic coupling in two marine apex predators. J. Anim. Ecol. 80, 294–305.

McConnaughey, T., McRoy, C.P., 1979. Food-web structure and the fractionation of carbon isotopes in the Bering Sea. Mar. Biol. 53, 257–262.

McCutchan, J.H., Lewis, W.M., Kendall, C., McGrath, C.C., 2003. Variation in trophic shift for stable isotope ratios of carbon, nitrogen, and sulfur. Oikos 102, 378–390.

McMahon, K.W., McCarthy, M.D., 2016. Embracing variability in amino acid $\delta^{15}N$ fractionation: mechanisms, implications, and applications for trophic ecology. Ecosphere 7, e01511.

Michner, R., Kaufman, L., 2005. Stable isotope ratios as tracers in marine food webs: an update. In: Micher, R., Kaufman, L. (Eds.), Stable Isotopes in Ecology and Environmental Science. Blackwell Publishing, Malden, MA, pp. 238–282.

Milessi, A.C., Danilo, C., Laura, R.G., Daniel, C., Javier, S., Rodríguez-Gallego, L., 2010. Trophic mass-balance model of a subtropical coastal lagoon, including a comparison with a stable isotope analysis of the food-web. Ecol. Model. 221, 2859–2869.

Miller, T.W., Brodeur, R.D., Rau, G.H., 2008. Carbon stable isotopes reveal relative contribution of shelf-slope production to the northern California Current pelagic community. Limnol. Oceanogr. 53, 1493–1503.

Miller, T.W., Brodeur, R.D., Rau, G., Omori, K., 2010. Prey dominance shapes trophic structure of the northern California Current pelagic food web: evidence from stable isotopes and diet analysis. Mar. Ecol. Prog. Ser. 420, 15–26.

Mongillo, T.M., Ylitalo, G.M., Rhodes, L.D., O'Neill, S.M., Noren, D.P., Hanson, M.B., 2016. Exposure to a Mixture of Toxic Chemicals: Implications for the Health of Endangered Southern Resident Killer Whales. U.S. Department of Commerce, Seattle, WA. NOAA Technical Memorandum NMFS-NWFSC-135.

Moore, J.W., Semmens, B.X., 2008. Incorporating uncertainty and prior information into stable isotope mixing models. Ecol. Lett. 11, 470–480.

Moriarty, P.E., Essington, T.E., Ward, E.J., 2016. A novel method to estimate prey contributions to predator diets. Can. J. Fish. Aquat. Sci. 74, 168–177.

O'Neill, S.M., Ylitalo, G.M., West, J.E., Bolton, J., Sloan, C.A., Krahn, M.M., 2006. Regional patterns of persistent organic pollutants in five Pacific salmon species (*Oncorhynchus* spp) and their contributions to contaminant levels in northern and southern resident killer whales (*Orcinus orca*). Southern Resident Killer Whale Symposium, Seattle, WA, pp. 38–42.

Parnell, A.C., Inger, R., Bearhop, S., Jackson, A.L., 2010. Source partitioning using stable isotopes: coping with too much variation. PLoS One 5, e9672.

Parnell, A.C., Phillips, D.L., Bearhop, S., Semmens, B.X., Ward, E.J., Moore, J.W., Jackson, A.L., Grey, J., Kelly, D.J., Inger, R., 2013. Bayesian stable isotope mixing models. Environmetrics 24, 387–399.

Peterson, B.J., Fry, B., 1987. Stable isotopes in ecosystem studies. Annu. Rev. Ecol. Syst. 18, 293–320.

Peterson, B.J., Howarth, R.W., 1987. Sulfur, carbon, and nitrogen isotopes used to trace organic matter flow in the salt marsh estuaries of Sapelo Island. Georgia. Limn. Oceanogr. 32, 1195–1213.

Phillips, D.L., 2001. Mixing models in analyses of diet using multiple stable isotopes: a critique. Oecologia 127, 166–170.

Phillips, D.L., Inger, R., Bearhop, S., Jackson, A.L., Moore, J.W., Parnell, A.C., Semmens, B.X., Ward, E.J., 2014. Best practices for use of stable isotope mixing models in food-web studies. Can. J. Zool. 92, 823–835.

Pinnegar, J.K., Jennings, S., O'Brien, C.M., Polunin, N.V.C., 2002. Long-term changes in the trophic level of the Celtic Sea fish community and fish market price distribution. J. Appl. Ecol. 39, 377–390.

Post, D.M., 2002. Using stable isotopes to estimate trophic position: models, methods, and assumptions. Ecology 83, 703–718.

Quevedo, M., Svanback, R., Eklov, P., 2009. Intrapopulation niche partitioning in a generalist predator limits food web connectivity. Ecology 90, 2263–2274.

Ramos, R., González-Solís, J., 2012. Trace me if you can: the use of intrinsic biogeochemical markers in marine top predators. Front. Ecol. Eviron. 10, 258–266.

Reum, J.C.P., 2011. Lipid correction model of carbon stable isotopes for a cosmopolitan predator, spiny dogfish *Squalus acanthias*. J. Fish Biol. 79, 2060–2066.

Reum, J.C.P., Essington, T.E., 2013. Spatial and seasonal variation in $\delta^{15}N$ and $\delta^{13}C$ values in a mesopredator shark, *Squalus suckleyi*, revealed through multitissue analyses. Mar. Biol. 160, 399–411.

Reum, J.C.P., Marshall, K.N., 2013. Evaluating $\delta^{15}N$-body size relationships across taxonomic levels using hierarchical models. Oecologia 173, 1159.

Reum, J.C.P., Jennings, S., Hunsicker, M.E., 2015. Implications of scaled $\delta^{15}N$ fractionation for community predator–prey body mass ratio estimates in size-structured food webs. J. Anim. Ecol. 84, 1618–1627.

Ruzicka, J.J., Brink, K.H., Gifford, D.J., Bahr, F., 2016. A physically coupled end-to-end model platform for coastal ecosystems: simulating the effects of climate change and changing upwelling characteristics on the Northern California Current ecosystem. Ecol. Model. 331, 86–99.

Samhouri, J.F., Levin, P.S., Harvey, C.J., 2009. Quantitative evaluation of marine ecosystem indicator performance using food web models. Ecosystems 12, 1283–1298.

Semmens, B.X., Ward, E.J., Moore, J.W., Darimont, C.T., 2009. Quantifying inter- and intra-population niche variability using hierarchical Bayesian stable isotope mixing models. PLoS One 4, e6187.

Shepherd, T.D., Myers, R.A., 2005. Direct and indirect fishery effects on small coastal elasmobranchs in the northern Gulf of Mexico. Ecol. Lett. 8, 1095–1104.

Simpfendorfer, C., Heupel, M., White, W., Dulvy, N., 2011. The importance of research and public opinion to conservation management of sharks and rays: a synthesis. Mar. Freshw. Res. 62, 518–527.

Stevens, J., 2000. The effects of fishing on sharks, rays, and chimaeras (chondrichthyans), and the implications for marine ecosystems. ICES J. Mar. Sci. 57, 476–494.

Stock, B.C., Semmens, B.X., 2016. Unifying error structures in commonly used biotracer mixing models. Ecology 97, 2562–2569.

Sweeting, C.J., Polunin, N.V.C., Jennings, S., 2006. Effects of chemical lipid extraction and arithmetic lipid correction on stable isotope ratios of fish tissues. Rapid Commun. Mass Spectrom. 20, 595–601.

Vanderklift, M.A., Ponsard, S., 2003. Sources of variation in consumer-diet $\delta^{15}N$ enrichment: a meta-analysis. Oecologia 136, 169–182.

Vaudo, J.J., Heithaus, M.R., 2011. Dietary niche overlap in a nearshore elasmobranch mesopredator community. Mar. Ecol. Prog. Ser. 425, 247–260.

Wai, T.C., Yeung, J.W.Y., Lam, V.Y.Y., Leung, K.M.Y., Dudgeon, D., Williams, G.A., 2012. Monsoons and habitat influence trophic pathways and the importance of terrestrial-marine linkages for estuary sharks. Ecosphere 3, 8.

Ward-Paige, C.A., Mora, C., Lotze, H.K., Pattengill-Semmens, C., McClenachan, L., Arias-Castro, E., Myers, R.A., 2010. Large-scale absence of sharks on reefs in the Greater-Caribbean: a footprint of human pressures. PLoS One 5, e11968.

Weinstein, M.P., Litvin, S.Y., Guida, V.G., 2009. Essential fish habitat and wetland restoration success: A tier III approach to the biochemical condition of common mummichog *Fundulus heteroclitus* in common reed *Phragmites australis*- and smooth cordgrass *Spartina alterniflora*-dominated salt marshes. Estuar. Coasts 32, 1011–1022.

Williams, G.D., Andrews, K.S., Farrer, D.A., Levin, P.S., 2010. Catch rates and biological characteristics of bluntnose sixgill sharks in Puget Sound. Trans. Am. Fish. Soc. 139, 108–116.

Williams, G.D., Harvey, C.J., Reum, J.C.P., Andrews, K.S., Levin, P.S. Trophic ecology and movement patterns of sixgill sharks in Puget Sound: a stable isotope analysis. Mar. Ecol. Prog. Ser. in review.

CHAPTER SIX

Age and Growth of Elasmobranchs and Applications to Fisheries Management and Conservation in the Northeast Pacific Ocean

Mary E. Matta*,[1], Cindy A. Tribuzio†, David A. Ebert‡, Kenneth J. Goldman§, Christopher M. Gburski*

*Resource Ecology and Fisheries Management, Alaska Fisheries Science Center, National Marine Fisheries Service, National Oceanic and Atmospheric Administration, Seattle, WA, United States

†Auke Bay Laboratories, Alaska Fisheries Science Center, National Marine Fisheries Service, National Oceanic and Atmospheric Administration, Juneau, AK, United States

‡Pacific Shark Research Center, Moss Landing Marine Laboratories, Moss Landing, CA, United States

§Alaska Department of Fish and Game, Homer, AK, United States

[1]Corresponding author: e-mail address: beth.matta@noaa.gov

Alaska skate (*Bathyraja parmifera*)

Contents

Advances in Marine Biology, Volume 77
ISSN 0065-2881
http://dx.doi.org/10.1016/bs.amb.2017.06.002

Abstract

In addition to being an academic endeavour, the practical purpose of conducting age and growth studies on fishes is to provide biological data to stock assessment scientists and fisheries managers so they may better understand population demographics and manage exploitation rates. Age and size data are used to build growth models, which are a critical component of stock assessments. Though age determination of elasmobranchs in the northeast Pacific Ocean (NEP) began in the 1930s, the field has evolved substantially in recent years, allowing scientists to incorporate age data into assessments for more species than ever before. Owing to the highly diverse biology of this group of fishes, each species has its own set of challenges with regard to age determination. Age determination methods typically rely on semicalcified hard structures that form regular growth patterns; however, the structure selected and preparation method used is often species specific. New staining techniques have improved the ability to assess age and improve ageing precision for some species, and advances in microchemical methods have allowed for independent means of estimating age and validating age determination accuracy. Here we describe current age determination methods for NEP elasmobranchs. While the library of available techniques is increasing, there are still some NEP species for which reliable ageing methods have yet to be defined; we discuss these challenges and potential avenues of future research. Finally, we conclude by describing how age estimates are used in growth models and subsequently in stock assessments of selected NEP elasmobranchs.

1. INTRODUCTION

Elasmobranchs are often characterized as equilibrium (K-selected) species because they typically have low fecundity, late maturity, high juvenile survivorship, slow growth rates, and high longevity (Camhi et al., 1998; Goldman et al., 2012; Hoenig and Gruber, 1990; King and McFarlane, 2003; Winemiller and Rose, 1992). These same life history characteristics can make them vulnerable to fishing mortality (Winemiller, 2005), especially when juveniles are removed before they are capable of reproduction and when adult females are removed at the height of their fecundity. Accurate age determination of individuals is an important factor for determining stock health. Age estimates are used in demographic and population dynamics models for stock assessments, which in turn are used to determine spawning stock biomass and set appropriate catch rates and limits. Additionally, growth-related life history parameters such as age-at-maturity, growth coefficients, and longevity estimates are related to the intrinsic rate of population growth and the instantaneous rate of natural mortality (Cortes, 2002; Hoenig, 1983; Jensen, 1996), highlighting the importance of obtaining reliable estimates of these parameters.

Elasmobranch skeletons lack bone and are primarily composed of calcified cartilage, but many species possess hard structures such as vertebrae, spines, and thorns that can be used to determine age. Age determination relies on the assumption that growth zones visible within these hard structures are deposited annually. However, age determination methods must be validated for accuracy by independent methods, because growth zones may not form continuously throughout the life of the fish or may not be related to temporal periodicity (Cailliet, 2015). Furthermore, there can be tremendous difficulties in obtaining age estimates from elasmobranchs for a number of reasons. One primary difficulty is that species differ in the degree of calcification present within their hard structures. Sometimes, multiple structures must be evaluated for their ability to form growth zone patterns and suitability for age determination. Certain species, such as Pacific Sleeper Sharks (*Somniosus pacificus*), do not possess hard structures that form readily visible growth patterns. Other species may be 'ageable' but are rarely encountered; thus, sample sizes are insufficient to assess the age structure of the population, validate age methods, or support stock assessments. Lastly, growth zones in hard structures can be faint in appearance or difficult to enumerate repeatedly, which can result in low precision of age estimates.

Before an age determination method can be considered viable for use in stock assessments, it should be validated. In other words, the accuracy of age estimates generated by the method must be supported, either directly or indirectly. Methods that confirm absolute age are considered the strongest form of validation; these are sometimes referred to as 'determinate' validation methods (Cailliet, 2015). Methods that assess the consistency or repeatability of an age determination method, or that only validate partial ages, have been alternately referred to as 'corroboration', 'verification', or 'indeterminate validation' methods (Cailliet, 2015; Cailliet et al., 2006; Goldman, 2005; Kimura et al., 2006). These methods are weaker but can be used to corroborate age estimates when determinate validation methods are unavailable (Kimura et al., 2006). All validation methods can be considered to exist along a continuum, varying in the degree to which they support the age determination criteria of a given species (Kimura et al., 2006). Below, we discuss several of the most commonly used forms of validation in shark age and growth studies. For a more complete discussion of age validation methods, see Campana (2001), Goldman et al. (2012), and Cailliet (2015) for excellent reviews.

One of the strongest and most commonly used forms of validation is bomb-derived radiocarbon (^{14}C) analysis (e.g., Campana et al., 2006). Elevated ^{14}C levels in hard structures are attributed to the 'bomb radiocarbon pulse', corresponding to a period of intense atmospheric testing of atomic bombs during the 1950s and 1960s. This atmospheric ^{14}C eventually mixed into oceanic water and has been demonstrated to impart a signal in the hard parts of marine animals, which can be used to date these materials with a high degree of accuracy (Campana, 2001; Goldman et al., 2012; Kalish, 1993). This method has been successfully used to validate age estimates derived from dorsal fin spines of Atlantic (*Squalus acanthias*) and North Pacific (*Squalus suckleyi*) Spiny Dogfish (Campana et al., 2006) and from vertebral centrum sections of Longnose Skate (*Raja rhina*) (King et al., 2017). In some species, including Porbeagle Shark (*Lamna nasus*) (Francis et al., 2007), Thorny Skate (*Amblyraja radiata*) (McPhie and Campana, 2009), Common Thresher Shark (*Alopias vulpinus*) (Natanson et al., 2016), Sand Tiger Shark (*Carcharias taurus*) (Passerotti et al., 2014), and White Shark (*Carcharodon carcharias*) (Andrews and Kerr, 2015; Hamady et al., 2014), bomb radiocarbon analysis has suggested that age estimates based on vertebral growth zone counts could be inaccurate, at least for older fish, with underestimation of true age occurring as somatic growth slows over time and growth zones either become increasingly compressed or stop accruing completely.

Another commonly used age validation method is the mark–recapture of fish chemically tagged with a calcium-binding compound such as oxytetracycline, OTC (e.g., McFarlane and Beamish, 1987). Mark–recapture studies were more common before the advent of bomb radiocarbon methods in the early 1990s but continue to be used today. The mark–recapture method typically does not validate absolute age, unless age is known at time of tagging; instead, it validates the number of growth zones deposited since tagging.

Thus, this method becomes more powerful the longer fish are at liberty (e.g., McFarlane and King, 2009).

Two of the most commonly employed indeterminate validation methods are edge analysis and marginal increment analysis, both of which attempt to verify the annual periodicity of growth zone formation. In edge analysis, the type of growth zone present on the structure's edge (opaque or translucent) can be plotted as a function of month or season of collection for numerous specimens of a species to determine if there is an annual pattern in growth zone formation (Cailliet, 2015; Goldman, 2005; Tribuzio et al., in press). Similarly, in marginal increment analysis, the ratio between the marginal growth zone width (i.e., growth that occurred during the collection year) and the penultimate growth zone width (i.e., growth that occurred during the previous year) can be graphed for multiple specimens according to the month of collection to determine whether there is an annual pattern of formation (Conrath et al., 2002; Goldman and Musick, 2006; Matta and Gunderson, 2007; Natanson et al., 1995). One of the difficulties associated with these methods is that to have adequate statistical power, they require samples to be collected throughout the year, which can be logistically infeasible, especially for species in far northern latitudes. Sometimes, it can be difficult to confidently interpret edge types or to obtain precise measurements for individuals with vague increment boundaries. These analyses can be further confounded by differences in accretion rates among age classes, or over different years or geographic regions (Kimura et al., 2007). Therefore, whenever possible, it is best to constrain these analyses to specimens collected from a single calendar year and region. Furthermore, due to the potential for age-related differences in growth rates, analyses should be repeated separately for each age group, and for as many age groups as possible. If these methods are not possible for older fish due to the greater tendency for growth zone compression and difficulty in edge interpretation with increasing age, the periodicity of growth zone formation can only be considered verified for the age groups assessed.

This chapter profiles age determination methods of northeast Pacific Ocean (NEP) elasmobranchs and describes how age and growth data are currently used in stock assessments of several key species in this region. While this does not represent an exhaustive list of all possible age determination methods for elasmobranchs (see Cailliet, 2015; Cailliet and Goldman, 2004; Goldman et al., 2012 for reviews), it is meant to describe the most commonly employed structure-based methods, highlighting previous work done specifically on NEP species. We refer to species occurring in four regions of the NEP: the US West Coast (WC); British Columbia, Canada (BC); the Gulf of Alaska (GOA); and the Bering Sea/Aleutian Islands (BSAI). These regions coincide with management areas in which stock assessments are conducted (see examples below and in Kacev et al., in press). We discuss challenges associated with the age determination of some NEP shark species and highlight several advances in age and growth technologies that may be used to generate age estimates to be applied in future academic investigations and stock assessments. Finally, we conclude with descriptions of how age and growth data have been directly incorporated into regional stock assessments of three common NEP elasmobranch species.

2. AGE DETERMINATION METHODS OF NEP ELASMOBRANCHS

A number of structures are used to age elasmobranchs in the NEP, including vertebrae, spines, thorns (in skates), and tooth plates. Presumed annual growth zones observed in these hard structures are typically referred to as 'annuli', derived from the Latin word for 'ring' (Panfili et al., 2002). An annulus consists of one opaque and one translucent growth zone (Fig. 1). These growth zones are so named due to their optical properties. For example, when viewing a vertebral centrum section through a microscope using transmitted light, opaque growth zones appear dark (because light cannot pass through), and translucent growth zones appear light. When viewed with reflected light, opaque growth zones appear light and translucent zones

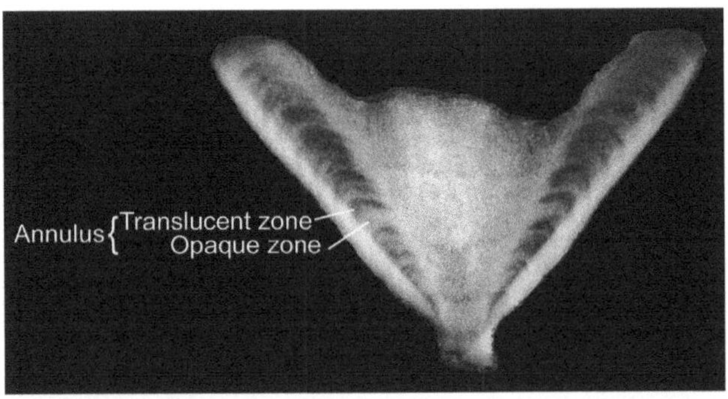

Fig. 1 Sectioned vertebral centrum from an Alaska Skate (*Bathyraja parmifera*), photographed with reflected light. One opaque growth zone and one translucent growth zone are assumed to equal 1 year of growth, or annulus. *Image has been enhanced using Adobe Photoshop Elements (v13.1) software.*

appear dark. Growth zone pairs, or annuli, are counted to arrive at an estimate of age. To enhance growth patterns, alternative methods such as staining are sometimes applied to hard structures. Image analysis and editing software such as Adobe Photoshop can also be used to enhance images and increase visual contrast between growth zones (Campana, 2014; Tribuzio et al., 2016b). Below is a review of structure-based age determination methods currently in use for NEP elasmobranchs.

2.1 Vertebra-Based Methods

2.1.1 Whole Vertebral Centra

Generally, NEP elasmobranchs are not aged from whole vertebral centra (Fig. 2A) due to the potential for underestimating age, because annuli on the outer portions of centra can be very tightly grouped, especially in older individuals (Cailliet, 1990; Cailliet et al., 1983). However, it may be possible to use whole centra to age very young, fast-growing individuals of certain species such as Blue Shark (*Prionace glauca*) (Campana, 2014).

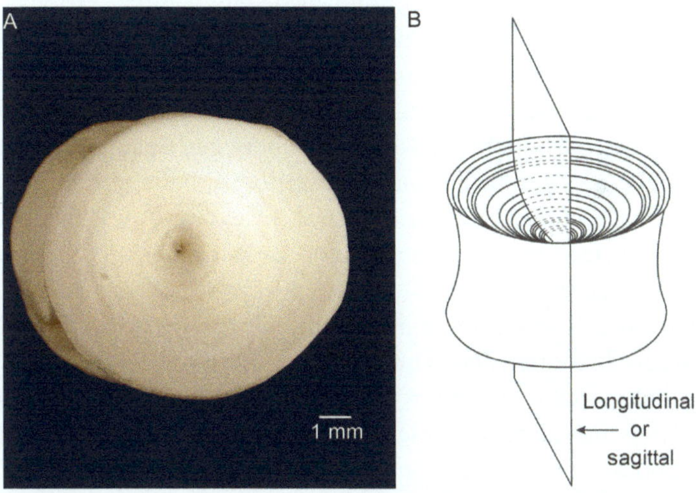

Fig. 2 Photograph (A) of a vertebral centrum from a Big Skate (*Beringraja binoculata*) viewed with reflected light, and illustration (B) of a typical elasmobranch vertebral centrum, showing preferred section plane. *Illustration adapted from Cailliet, G., Martin, L.K., Harvey, J.T., Kusher, D., Welden, B.A., 1983. Preliminary studies on the age and growth of Blue,* Prionace glauca, *Common Thresher,* Alopias vulpinus *and Shortfin Mako,* Isurus oxyrinchus, *Sharks from California waters, in: Prince, E.D., Pulos, L.M., (Eds.), Proceedings of the International Workshop on Age Determination of Oceanic Pelagic Fishes: Tunas, Billfishes, and Sharks. NOAA Tech. Rep. 8 by W. Carlson (AFSC).*

For this method, a length of vertebral column containing 8–12 centra is removed anterior to the first dorsal fin and frozen. Vertebral columns are thawed, and soft tissue and transverse processes are removed using a sharp knife. Individual centra are soaked in distilled water for 5 min and air-dried. Diluted bleach (10%) can be used for larger centra, with soak times from 5 to 30 min. If desired, the centra may be soaked in formic acid for 2–4 min to remove any remaining bleach and tissue. Whole centra may be stained with silver nitrate (Cailliet et al., 1983; Stevens, 1975) or subjected to graphite microtopography (Neer and Cailliet, 2001; Parsons, 1983) and X-radiography (Cailliet et al., 1983) to enhance contrast between growth zones.

2.1.2 Centrum Sections

The method of age determination that is the most common in the elasmobranch age and growth literature and widely used on NEP species is sagittal thin sectioning of vertebral centra. In this method, thoracic vertebrae are cleaned of transverse processes using a sharp knife, soaked in 70% ethanol, and air dried. Vertebral centra are then typically embedded in resin to facilitate sectioning, although centra from larger individuals can be sectioned without embedding. Using a diamond-blade saw, sagittal sections (Fig. 2B) are cut from centra, mounted onto glass slides, and, optionally, wet sanded to remove saw blade marks and smooth the sample surface. Optimum section thickness is species dependent but can range from 0.3 to 1.0 mm. Water or ethanol can be applied to improve contrast between opaque and translucent growth zones, and either transmitted or reflected light is used for viewing, depending on section thickness. Mineral oil can also be applied but may result in over clearing of growth zones (Campana, 2014). The four 'arms' of the corpus calcareum (Fig. 3) are the primary reading surfaces. The age assessment process starts by counting annuli near the focus of the section and moves to the outer edge. The birthmark increment (age-0), a useful reference point, is often evident as an angle change in the corpus calcareum, although it is not always visible in all species (Goldman, 2005) (Fig. 3). Growth zones are followed from one arm of the corpus calcareum across the intermedialia to the other side if possible; this can be especially useful for identifying early annuli (Fig. 3).

Several studies have verified annual periodicity of growth zone formation in vertebral centrum sections using marginal increment analysis and edge analysis, although these attempts have been far more successful for species from regions other than the NEP (e.g., Conrath et al., 2002; Smith et al., 2007; Sulikowski et al., 2003, 2005), where results have often been

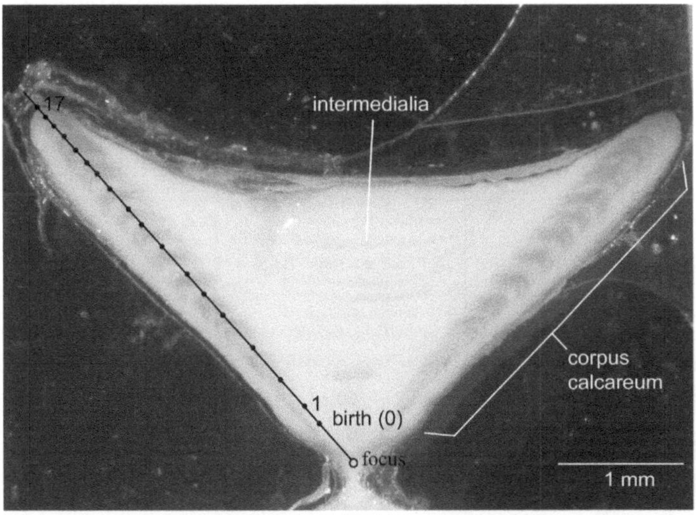

Fig. 3 Half of a sectioned vertebral centrum from a Longnose Skate (*Raja rhina*), showing two of the 'arms' of the corpus calcareum, photographed with reflected light, showing birthmark increment, intermedialia, and corpus calcareum. Translucent growth zones (dark growth zones, represented by *black dots*) were counted to arrive at an age estimate of 17 years.

inconclusive (Table 1). However, edge and marginal increment analyses on centrum sections have suggested annual periodicity of growth zone formation for Alaska Skate (*Bathyraja parmifera*) in the eastern Bering Sea (Matta and Gunderson, 2007) and Salmon Shark (*Lamna ditropis*) in the NEP (Goldman and Musick, 2006). This age determination method has been validated by bomb radiocarbon analysis for Longnose Skate (King et al., 2017).

2.1.3 Stained Centrum Sections

In this method, vertebral centra are sectioned and stained to enhance contrast between growth zones (Fig. 4). Sections can be created either using the protocol for unstained sections, described earlier, or they can be created employing histological techniques. Specifically, centra are embedded in paraffin and cut using a microtome to create very thin (<0.1 mm) sections (Natanson et al., 2007). Sections produced in this way can be up to an order of magnitude thinner than those created using a diamond saw blade. Optimum section thickness can vary according to species. However, care should be taken when interpreting very thin sections, as reduced section thickness could potentially lead the reader to interpret 'checks', or false annuli, as true annuli because checks are often more easily observed in thinner sections.

Table 1 Age-Related Life History Parameters for Northeast Pacific Elasmobranch Species

Species Name	Common Name	Region	Sex	T_{max}	L_∞	κ	t_0	Growth Models	Method	Validation	Reference
Hydrolagus colliei	Spotted Ratfish	British Columbia	M	16	82.4 PCL	0.03	−1.46	3VB, 2VB, 3G, L*	V	(EA)	1
			F	21	82.1 PCL	0.04	−0.26				
Hexanchus griseus	Bluntnose Sixgill Shark	British Columbia	—	—	—	—	—	—	N	—	2
Squalus suckleyi	North Pacific Spiny Dogfish	NE Pacific	—	52	—	—	—	—	F	BC	3
		British Columbia	—	70	—	—	—	—	F	OTC	4
		British Columbia	F	81	114.9 TLext	0.04	−3.56	3VB	F	—	5
		Gulf of Alaska	M	50	88.3 PCL	0.10	−4.80	3VB, 2VB, 2VB*, 2G, 3G, L	F	—	6
			F		121.4 PCL	0.03	−12.10				
Mustelus henlei	Brown Smoothhound	California	M	7	86.1 TL	0.29	−1.09	3VB	Other	(EA)	7
			F	13	97.6 TL	0.23	−1.38				
Triakis semifasciata	Leopard Shark	California	M		149.9 TL	0.09	−2.03	3VB	S	EA, OTC	8
			F		160.2 TL	0.07	−2.74				
Alopias vulpinus	Common Thresher Shark	California	M	15	492.7 TL	0.22	−1.42	3VB	W	—	9
			F		636.0 TL	0.16	−1.02				
			M		416.2 TL	0.19	−2.08	3VB	W		10
			F		464.3 TL	0.12	−3.35				

Prionace glauca	Blue Shark	California	M	9	295.3 TL	0.18	−1.11	3VB	W	—	9
			F		241.9 TL	0.25	−0.80				
		North Pacific	M		369.0 TL	0.10	−1.38	3VB	S	—	11
			F		304.0 TL	0.16	−1.01				
			M	15	289.7 PCL	0.13	−0.76	3VB	W	—	12
			F	15	243.3 PCL	0.14	−0.85				
Lamna ditropis	Salmon Shark	NE Pacific	M	17	182.8 PCL	0.23	−1.90	3VB	U	MIA	13
			F	20	207.4 PCL	0.17	−2.30				
Torpedo californica	Pacific Electric Ray	California	M	14	92.1 TL	0.14	−1.48	3VB*, 3G, L, N	W		14
			F	16	137.3 TL	0.07	−1.93				
Beringraja binoculata	Big Skate	Gulf of Alaska	M	15	153.3 TL	0.15	−0.63	3VB	U	—	15
			F	14	247.5 TL	0.08	−1.08				
		California	M	11	—	—	—	3VB, L*	U	EA	16
			F	12	—	—	—				
		British Columbia	M	25	233.0 TL	0.05	−2.10	3VB, L*	U	—	17
			F	26	293.5 TL	0.04	−1.60				

Continued

Table 1 Age-Related Life History Parameters for Northeast Pacific Elasmobranch Species—cont'd

Species Name	Common Name	Region	Sex	T_{max}	L_{∞}	κ	t_0	Growth Models	Method	Validation	Reference
Raja rhina	Longnose Skate	Gulf of Alaska	M	25	168.8 TL	0.06	−1.67	3VB	U	BC	15, 18
			F	24	234.1 TL	0.04	−1.99				
		California	M	13	96.7 TL	0.25	0.73	3VB*, L	U	(EA)	16
			F	12	106.9 TL	0.16	−0.30				
		British Columbia	M	23	131.5 TL	0.07	−2.17	3VB, L	U	—	17
			F	26	137.2 TL	0.06	−1.80				
		US West Coast	M	20	—	—	—	3VB, 2VB*, L	S	(MIA), EA	19
			F	22	—	—	—				
Bathyraja abyssicola	Deepsea Skate	US West Coast	F	45	—	—	—	N	S, (U), (C)	—	20
Bathyraja aleutica	Aleutian Skate	Bering Sea	M	16	170.5 TL	0.11	1.69	3VB*, 2VB, 3G, L	U	(MIA), (EA)	21
			F	17	174.4 TL	0.10	1.86				
		Gulf of Alaska	M	18	158.3 TL	0.09	3.41	3VB, 2VB*, 3G, L	U	(MIA), (EA)	21
			F	19	211.8 TL	0.05	4.55				
Bathyraja interrupta	Bering Skate	Bering Sea	M	18(S) 18(U)	126.3 TL(U)	0.05(U)	−3.26(U)	3VB, 2VB, 3G*, L*	U, S	—	22
			F	21(S) 19(U)	104.9 TL(U)	0.07(U)	−2.54(U)				
		Gulf of Alaska	M	21(U)	119.4 TL(U)	0.05(U)	−3.29(U)	3VB, 2VB, 3G*, L*	U, S	—	22
			F	21(U)	114.4 TL(U)	0.06(U)	−2.54(U)				

Bathyraja kincaidii	Sandpaper Skate	Central California	M	18	58.0 TL	0.19	−2.53	3VB*, 3G, L	U, (C)	(MIA), (EA)	23
			F	17	53.7 TL	0.24	−1.63				
Bathyraja lindbergi	Commander Skate	Bering Sea	M	35(S) 27(U)	119.3 TL(S) 132.0 TL(U)	0.05(S) 0.04(U)	−2.63(S) −4.27(U)	3VB, 2VB, 3G*, L*	U, S, (C)	(MIA), (EA)	24
			F	32(S) 32(U)							
Bathyraja maculata	Whiteblotched Skate	Bering Sea	M	36(S) 32(U)	169.6 TL(S) 155.6 TL(U)	0.03(S) 0.04(U)	−5.41(S) −4.25(U)	3VB, 2VB, 3G*, L*	U, S, (C)	(MIA), (EA)	24
			F	33(S) 32(U)							
Bathyraja minispinosa	Whitebrow Skate	Bering Sea	M	35	129.0 TL	0.03	−4.86	3VB, 2VB, 3G*, L, Richards, polynomial	U, (C)	(MIA), (EA)	25
			F	37	171.9 TL	0.02	−6.66				
Bathyraja parmifera	Alaska Skate	Bering Sea	M	15(U)	126.3 TL(U)	0.12(U)	−1.39(U)	3VB, 3G*	U, C	MIA, EA	26
			F	17(U)	144.6 TL(U)	0.09(U)	−1.75(U)				
			F	26(S) 24(U)					U, S	—	27
Bathyraja trachura	Roughtail Skate	Bering Sea	M	35	119 TL	0.04	−3.03	3VB, G, L*, S	S	—	28
			F	36							
		US West Coast	M	20	112.1 TL	0.06	−3.45	3VB, 2VB*, 3G	U, (C)	(MIA), (EA)	29
			F	17							

Continued

Table 1 Age-Related Life History Parameters for Northeast Pacific Elasmobranch Species—cont'd

Species Name	Common Name	Region	Sex	T_{max}	L_∞	κ	t_0	Growth Models	Method	Validation	Reference
Bathyraja stellulata	Starry Skate	California	M	14(S) 10(U)	96.9 TL(S) 99.3 TL(U)	0.08(S) 0.11(U)	−1.67(S) −1.51(U)	3VB, 2VB*, 3G, L	U, S, (C)	(MIA), (EA)	30
			F	15(S) 11(U)	116.2 TL(S) 83.6 TL(U)	0.05(S) 0.16(U)	−2.78(S) −0.88(U)				

M = male, F = female. Maximum observed age is T_{max} (year). For studies that tested multiple growth models (3VB = three-parameter von Bertalanffy, 2VB = two-parameter von Bertalanffy, 2PVB = two-phase von Bertalanffy, 2G = two-parameter Gompertz, 3 = 3-parameter Gompertz, L = logistic, N = linear, S = general form of the Schnute model), the authors' preferred model is indicated by an *asterisk*. For the purpose of comparison among species, only 3VB model parameters (maximum theoretical length L_∞, in cm total length TL, extended total length TL_{ext}, or precaudal length PCL; growth coefficient κ, year^{-1}, theoretical age at zero length t_0, year) are shown (when available). Methods used to determine age estimates are shown are designated as C = caudal thorns, F = dorsal fin spines, N = neural arches, S = stained sections of vertebral centra, U = unstained sections of vertebral centra, V = vomerine tooth plates, and W = whole vertebral centra. When available, age validation techniques (BC = bomb radiocarbon, OTC = oxytetracyline, MIA = marginal increment analysis, EA = edge analysis) are described. Age determination methods and validation methods that were incomplete, inconclusive, or unsuccessful are shown in *parentheses*.

[1]King and McPhie (2015), [2]McFarlane et al. (2002), [3]Campana et al. (2006), [4]Beamish and McFarlane (1985), [5]Saunders and McFarlane (1993), [6]Tribuzio et al. (2010), [7]Yudin and Cailliet (1990), [8]Kusher et al. (1992), [9]Cailliet et al. (1983), [10]Smith et al. (2008), [11]Tanaka et al. (1990), [12]Nakano (1994), [13]Goldman and Musick (2006), [14]Neer and Cailliet (2001), [15]Gburski et al. (2007), [16]Zeiner and Wolf (1993), [17]McFarlane and King (2006), [18]King et al. (2017), [19]Thompson (2005), [20]Provost (2016), [21]Haas et al. (2016), [22]Ainsley et al. (2014), [23]Perez et al. (2011), [24]Maurer (2009), [25]Ainsley et al. (2011), [26]Matta and Gunderson (2007), [27]Matta (AFSC, unpublished data), [28]Winton et al. (2014), [29]Davis et al. (2007), and [30]James et al. (2014).

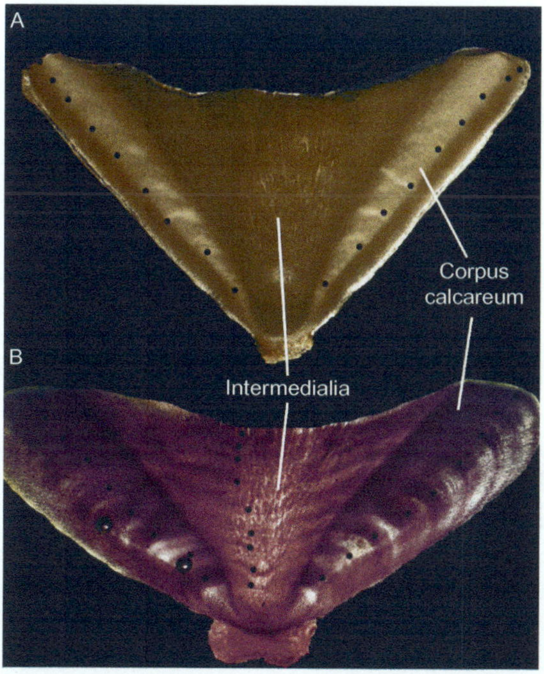

Fig. 4 Unstained (A) and stained (B) and sectioned vertebral centra from an Alaska Skate (*Bathyraja parmifera*) estimated to be 9 years old. Where possible, growth zones are followed from the intermedialia (central portion of the vertebra) to the corpus calcareum to arrive at an age estimate. Counted growth zones are indicated by *black dots. Photographs were taken with transmitted light. Images have been enhanced using Adobe Photoshop Elements (v13.1) software.*

In addition to sectioning, staining is employed to improve the appearance of growth zones. A histological procedure in which centra are decalcified and stained with haematoxylin, then sectioned to <0.1 mm (Natanson et al., 2007) has been applied to several NEP species (Ainsley et al., 2014; Maurer, 2009; Winton et al., 2014). Other stains employed in clarifying sections are alizarin red, crystal violet, silver nitrate, and cobalt nitrate (Cailliet and Goldman, 2004; Cailliet et al., 1983; Goldman et al., 2012; Hoenig and Brown, 1988; Thompson, 2005).

Recently, a modified haematoxylin-staining technique (Bubley et al., 2012) was applied to Alaska Skate centrum sections of 0.4 mm thickness to clarify banding patterns and improve age estimate precision (Fig. 4). Generally, ages were in agreement between stained and unstained thin sections, with similar maximum observed ages (Table 1), and there was no systematic bias to suggest that the methods resulted in different ages

(M.E. Matta, Alaska Fisheries Science Center AFSC NOAA, unpublished data). Overall, interreader bias was reduced and interreader precision improved using the staining technique for Alaska skate, although, as is typical for most elasmobranchs, precision was still generally low due to the relative difficulty of interpreting age structures. While staining generally seems to improve growth zone pattern clarity, this method awaits direct validation. Bubley et al. (2012) used marginal increment analysis to verify annual periodicity of growth zone formation in stained centrum sections for Atlantic Spiny Dogfish. A similar study on North Pacific Spiny Dogfish (*S. suckleyi*) also verified the annual periodicity of annuli in stained centrum sections of young fish but found a bias between this method and the traditional method of ageing, dorsal fin spines (described later), for older fish (Tribuzio et al., in press).

Some studies have reported a bias between unstained 'gross' (0.3–0.4 mm) sections and histological thin (0.08–0.10 mm) sections. Histological sections produced consistently greater annulus counts than gross sections for Commander (*Bathyraja lindbergi*) and Whiteblotched Skates (*Bathyraja maculata*) (Maurer, 2009), Bering Skate (*Bathyraja interrupta*) (Ainsley et al., 2014), Starry Skate (*Raja stellulata*) (James et al., 2014), and Roughtail Skate (*Bathyraja trachura*) (Winton, 2011). The authors of these studies attributed the differences to greater difficulty visualizing growth zones in gross sections, which was reflected by lower precision between age estimates for gross sections than for histological sections. When possible, multiple methods should be tested for a given species, with the ultimate goal of validating age estimates, as such differences in age determination methodology can have large impacts on growth models and life history parameters (Table 1) (Maurer, 2009).

2.1.4 Neural Arches

In species where centra are poorly calcified and other hard structures are not available, neural arches could potentially offer an alternative ageing structure (Fig. 5). Neural arches, which enclose the spinal cord, are located on the dorsal surface of vertebral centra. To prepare neural arches for age determination, first a section of vertebral column (15–20 vertebrae), including the neural and haemal arches, is removed just posterior to the head. Vertebrae are cleaned, soaked in bleach for 15 min, and rinsed in distilled water. Neural arches are soaked in a 1% solution of silver nitrate stain for a minimum of 1 h while exposed to wide-spectrum light (320–400 nm). After rinsing in distilled water, dorsoventral sections are cut at a thickness of 2–3 mm. Sections are placed under a dissecting microscope using white light (400–700 nm) to

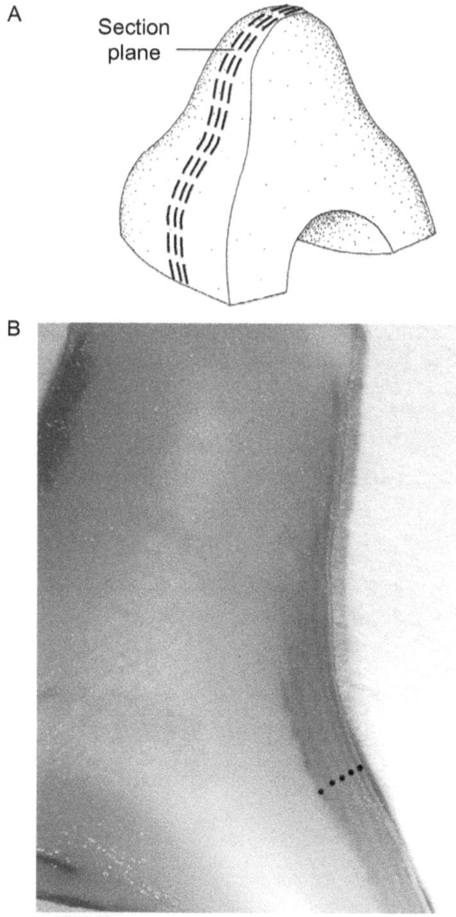

Fig. 5 Schematic illustration of a neural arch (A) and photograph of a sectioned and stained neural arch (B) from a Bluntnose Sixgill Shark (*Hexanchus griseus*). Potential growth zones are indicated by *black dots. Reproduced with permission from McFarlane, G.A., King, J.R., Saunders, M.W., 2002. Preliminary study on the use of neural arches in the age determination of Bluntnose Sixgill Sharks (*Hexanchus griseus*). Fish. Bull. 100, 861–864.*

allow growth bands to appear. The development process is halted by soaking the sections in a solution of 1.0 L water and 125 mL 28% acetic acid. Stained neural arch sections are microscope slide mounted with an acrylic toluene mounting medium (McFarlane et al., 2002).

Neural arches have been used to generate preliminary age estimates for Bluntnose Sixgill Sharks (*Hexanchus griseus*) in the NEP (McFarlane et al., 2002), though it should be noted that the observed banding patterns in neural arches have not been verified as annual. Furthermore, this method has not been successfully applied to any other elasmobranch species to date. Therefore, caution should be used when applying this method, and attempts should be made at age validation.

2.2 Caudal Thorns

Certain species of skate possess thorns along the midline of their tails (Fig. 6) which can used for age determination (e.g., Gallagher and Nolan, 1999). These caudal thorns can be sectioned, although they are primarily read whole; stains such as silver nitrate are also sometimes applied. Thorns are removed from the most anterior section of the tail, being careful to include the thorn base (Fig. 7A). They are cleaned by boiling them in tap water for 10–20 min, and excess soft tissue is wiped away with forceps and laboratory tissue. To clarify the banding pattern, thorns may be stained with silver nitrate. Thorns are placed against a dark background and aged using a dissecting microscope with fibre-optic reflected light, and are sometimes simultaneously viewed with transmitted light. As with vertebrae, thorns have opaque and translucent growth zones; the translucent growth zones have a ridge-like appearance, which can be enhanced by adjusting the angle of the reflected light source to allow for easier interpretation of the growth pattern (Fig. 7B). The ridges are counted in reference to the protothorn, or year-0, analogous to the 'birthmark' increment observed in vertebral centrum thin sections (Fig. 7B).

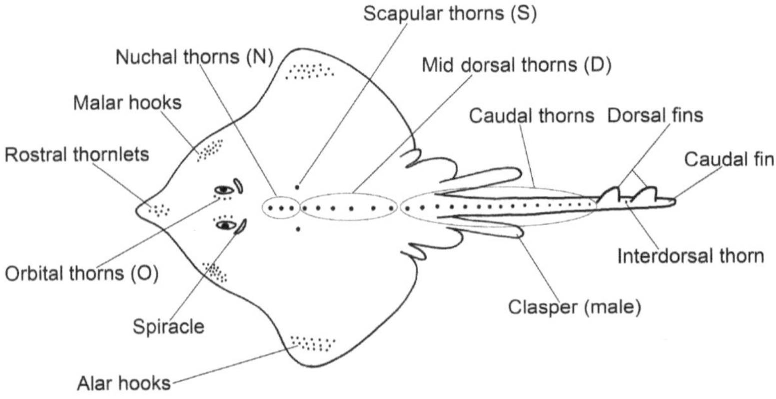

Fig. 6 Diagram of a generalized skate, showing caudal thorns. Caudal thorns used for age determination are typically those located most anteriorly, in the region of the pelvic fins. *Illustration adapted by D. Stevenson (AFSC) for this manuscript from Stevenson, D.E., Orr, J.W., Hoff, G.R., McEachran, J.D., 2007. Field Guide to Sharks, Skates, and Ratfish of Alaska. Alaska Sea Grant College Program, University of Alaska Fairbanks, Fairbanks, Alaska.*

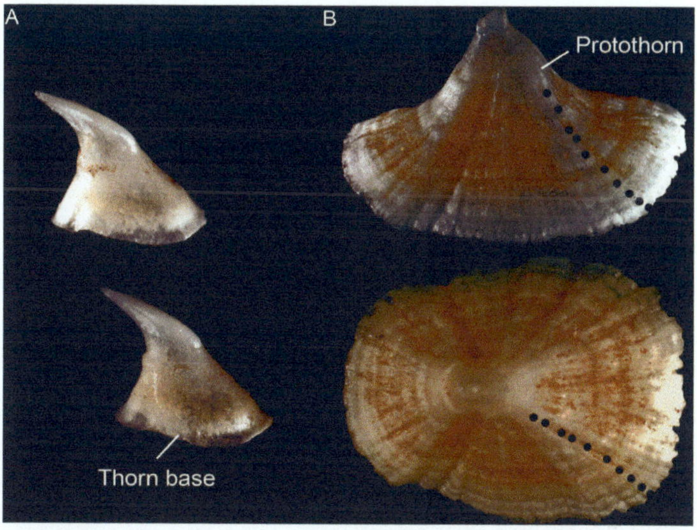

Fig. 7 Caudal thorns, photographed with reflected light, from an age-0 Alaska Skate (*Bathyraja parmifera*) (A) and an Alaska Skate estimated to be 10 years (B). Protothorn, or 'birthmark increment' (age-0), and growth zones (represented by *black dots*) are indicated in (B). *Images in (A) and (B) are not on the same scale, and were enhanced using Adobe Photoshop Elements (v13.1) software.*

The caudal thorn method was first introduced by Gallagher and Nolan (1999) as a less laborious and potentially nonlethal alternative age determination method. This method has produced similar age estimates to vertebral centrum sections for several species of Atlantic skate (Gallagher and Nolan, 1999; Gallagher et al., 2006; Henderson et al., 2004; Serra-Pereira et al., 2008). A number of studies have attempted to use caudal thorns for age determination of NEP species (Table 1 for published studies; D.A. Ebert, unpublished data, Pacific Shark Research Center, PSRC, Moss Landing Marine Laboratories MLML); however, age estimates from thorns and vertebral centra have only agreed for one species in this region, the Alaska Skate (Matta and Gunderson, 2007). One disadvantage of caudal thorns is that they are located on the exterior of the body and are thus subject to wear. The caudal thorn method has been verified using edge analysis for Thornback Ray (*Raja clavata*) collected off the coast of Portugal (Serra-Pereira et al., 2008), but this method has not yet been validated for any NEP species.

2.3 Dorsal Fin Spines

Dorsal fin spines have been successfully used to age one elasmobranch species in the NEP, the North Pacific Spiny Dogfish. The Spotted Ratfish

(*Hydrolagus colliei*), which is not a true elasmobranch but a member of the chondrichthyan class and a close relative of sharks, also possesses dorsal fin spines. However, work by Barnett et al. (2009) using microcomputed tomography suggested that while banding patterns are visible in Spotted Ratfish dorsal fin spines, they do not have mineral density gradients and thus likely do not exhibit consistent temporal periodicity.

Recently, the fin-spine method applied to North Pacific Spiny Dogfish was reexamined by Tribuzio et al. (in press), with a detailed description of the method in Tribuzio et al. (2016b). In short, the spines are separated from the vertebral column, ensuring that the enamel is not damaged. They are then cleaned of soft tissue, dried, and stored in paper envelopes. Annuli are counted by examining the spine under a dissecting microscope with reflected light (Fig. 8A). Growth zones are not considered opaque or translucent, but are referred to as 'dark' or 'light' because of how the enamel is deposited on the surface of the spine.

Spines are exposed to the environment and thus are often broken or worn (Fig. 8B), necessitating estimating the number of lost annuli. A method

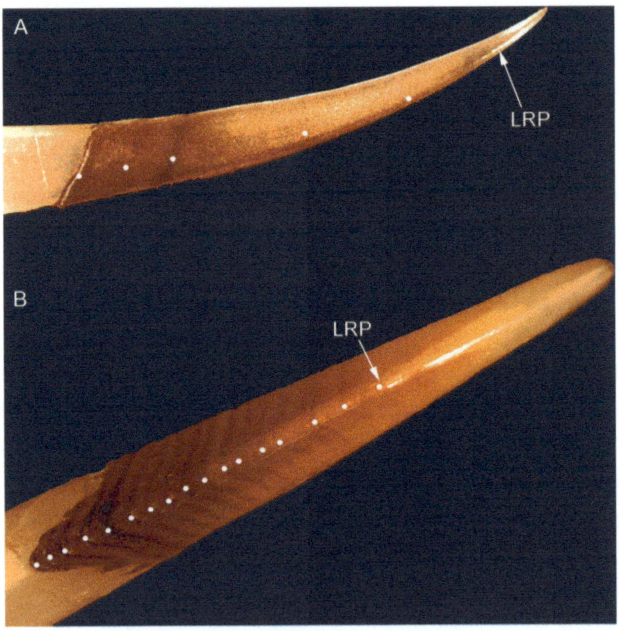

Fig. 8 North Pacific Spiny Dogfish (*Squalus suckleyi*) dorsal fin spines photographed with reflected light, showing examples of little wear (A) and significant wear (B). LRP is the last readable point (immediately preceding wearing of the spine enamel). Counted growth zones are indicated by *white dots. Image has been enhanced using Adobe Photoshop Elements (v13.1) software.*

proposed by Ketchen (1975) that is still in use today applies a simple exponential relationship between the base diameter of the spine and the diameter of the spine at the last readable point (LRP). The LRP corresponds to the point on the spine immediately preceding enamel erosion (Fig. 8B). Annuli are counted between these two landmarks on the spine, and any 'lost' annuli distal to the LRP are estimated and added to the count. Dorsal fin spine age estimates have been validated for North Pacific Spiny Dogfish by two direct methods: OTC-injected fish which were tagged, released, and subsequently recaptured up to 20 years later (Beamish and McFarlane, 1985; McFarlane and Beamish, 1987; McFarlane and King, 2009), and bomb-derived radiocarbon (Campana et al., 2006) (Fig. 9).

2.4 Vomerine Tooth Plates

Because the Spotted Ratfish lacks suitable ageing structures typically used for age and growth estimates, an alternative method was developed by counting tritor ridges on the vomerine tooth plate (King and McPhie, 2015). These ridges are easy to identify and may be an effective structure for use in estimating age for the Spotted Ratfish (Fig. 10A), though the method

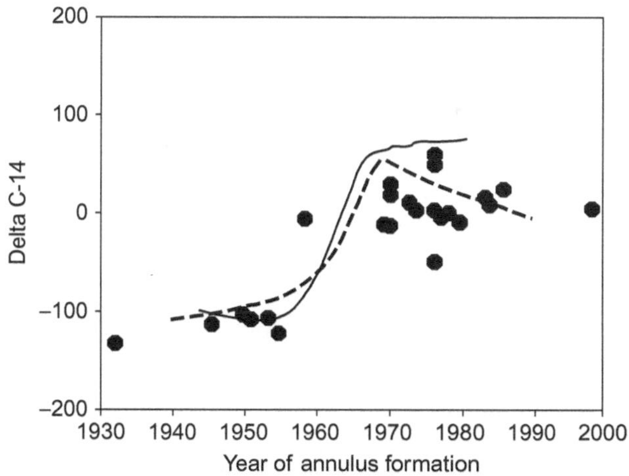

Fig. 9 Bomb-derived radiocarbon ($\Delta^{14}C$) in North Pacific Spiny Dogfish (*Squalus suckleyi*) dorsal fin spines, with year of formation based on presumed annual growth zone counts from spines. Measured values of dogfish spine $\Delta^{14}C$ (*black dots*) are compared to reference chronologies for Pacific Halibut (*Hippoglossus stenolepis; solid line*) and Yelloweye Rockfish (*Sebastes ruberrimus; dashed line*) otoliths from Piner and Wischniowski (2004) (*solid line*) and Kerr et al. (2004), respectively. *Reproduced with permission from Campana, S.E., Jones, C., McFarlane, G.A., Myklevoll, S., 2006. Bomb dating and age validation using the spines of Spiny Dogfish (*Squalus acanthias*). Environ. Biol. Fish. 77, 327–336.*

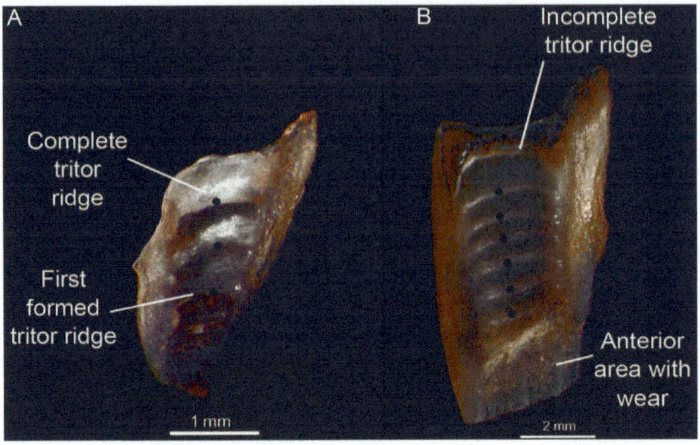

Fig. 10 Vomerine tooth plates from two different Spotted Ratfish (*Hydrolagus colliei*), photographed with reflected light. Tritor ridges (*black dots*) are indicated on a tooth plate with no wear (A) and a tooth plate with wear at the anterior edge (B). *Reproduced with permission from King, J.R., McPhie, R.P., 2015. Preliminary age, growth and maturity estimates of Spotted Ratfish (*Hydrolagus colliei*) in British Columbia. Deep-Sea Res. 115 (Pt. II), 55–63.*

has not been validated. To prepare the structures for age determination, the frozen jaws are thawed and tooth plates are separated and removed from the surrounding tissue using a scalpel and forceps. For smaller ratfish, tooth plates are removed under a dissecting microscope. Tooth plates are soaked in warm water 2–3 min to remove remaining tissue and placed in individual envelopes to dry (24–48 h). Tritor ridges are counted on either the left or right vomerine tooth plate. Ridges appear as elevated white bands under reflected light, and are counted if they extend from each side of the tooth plate and are distinct along its entire length (King and McPhie, 2015). Tritor ridges are subject to erosion (Fig. 10B), and therefore age may be underestimated in older individuals if corrections for wear are not made (King and McPhie, 2015).

3. CHALLENGES IN NEP ELASMOBRANCH AGE DETERMINATION AND POTENTIAL FOR FUTURE STUDIES

While age determination has been successful for some NEP species such as the Longnose Skate, age determination of others has not been without its challenges. Certain species cannot be aged from any of the hard structures described in the previous section. Here, we discuss several NEP species for which traditional age determination methods have not been effective. We also touch on several technological advances, many of which were first applied to the otoliths of teleost fishes, which may be worth exploring to estimate age and growth of these challenging NEP elasmobranch species. However, it should be noted that there are additional methods not mentioned here that could also be applicable to elasmobranch age determination, but for which application has not been tested to date.

Many attempts have been made to age Pacific Sleeper Shark (Wischniowski, 2008; K.J. Goldman, unpublished data; AFSC, unpublished data; D.A. Ebert, unpublished data), although none have been successful to date. In general, little is known regarding the biology, movements, abundance, or life history of this deep-water shark. Its vertebral column is poorly calcified, and individual vertebrae are not well articulated. At the AFSC, histological methods were employed to examine tissue microstructure and determine whether growth zones could be detected in vertebral centra, neural arches, and jaws (C.M. Gburski and M.E. Matta, AFSC NOAA, unpublished data). Structures were prepared by first embedding in paraffin and then sectioning to thicknesses ranging from 7 to 55 μm using a microtome. However, the microtome often shredded these very thin sections; therefore, thicker sections up to 6 mm were cut using a mandolin slicer. A variety of stains, including collagen-binding stains (1% aniline blue, Mallory's aniline blue), calcium-binding stains (1% neutral red, 2% alizarin red, von Kossa silver), and a collagen/nuclear stain (Harris modified haematoxylin and eosin) were applied to sections. Sections soaked in von Kossa silver stain were fluoresced under UV light for up to 2 h; all other sections were viewed using reflected or transmitted light, with and without a polarizer. Collagen fibre bundles in vertebral centrum sections were made apparent by the aniline blue stain, and somewhat resembled growth banding patterns (Fig. 11); however, we were not able to conclude whether these bundles were deposited on an annual basis. It is possible that they are related to growth of the organism and not consistently correlated with time. Chondrocytes (cartilage-secreting cells) were also observed within the collagen matrix. A concurrent study at the International Pacific Halibut Commission applied the stains crystal violet and alizarin red S to thick sections (5 mm) of vertebral centra, but did not discern any evidence of growth zones or microstructure (Wischniowski, 2008). In most sharks, vertebral centrum cartilage is well calcified and primarily areolar, while the skull and fin components are tessellated (Dean and Summers, 2006). Pacific Sleeper Shark vertebrae appear to be composed of prismatic or globular cartilage rather than areolar

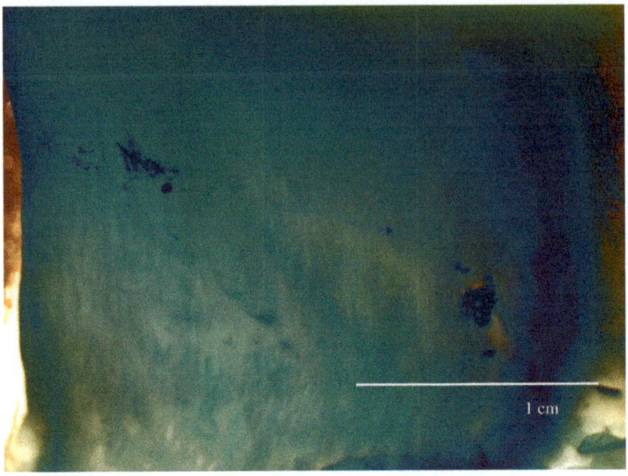

Fig. 11 Pacific Sleeper Shark (*Somniosus pacificus*) vertebral centrum sagittally sectioned (2–6 mm) and stained with 1% aniline blue. Collagen bundles appear in linear rows and somewhat resemble growth bands but cannot be assumed to be secreted on an annual basis. *Image taken with transmitted light and enhanced using Adobe Photoshop Elements (v13.1) software.*

cartilage, all of which are described in Dean and Summers (2006). Poor calcification seems to be a common feature in the vertebral centra of species in the shark order Hexanchiformes, including Broadnose Sevengill Shark (*Notorynchus cepedianus*) and Bluntnose Sixgill Shark, and of deep-water elasmobranchs such as Prickly Shark (*Echinorhinus cookei*), Longnose Catshark (*Apristurus kampae*), Brown Cat Shark (*Apristurus brunneus*), and Filetail Catshark (*Parmaturus xaniurus*) in the NEP (Braccini et al., 2010; Cailliet, 1990, 2015; Wischniowski, 2008, D.A. Ebert PSRC, MLML, unpublished data). For these species, alternative methods of age determination need to be explored.

A study on the Bluntnose Sixgill Shark tested pectoral girdles and mesoptergia as nontraditional ageing structures (Campbell, 2010). Stains (alizarin red S and silver nitrate) were applied to these structures and to neural arches in an attempt to reproduce results by McFarlane et al. (2002). Some banding patterns were detected using silver nitrate, although results were not consistent among specimens or structures (Campbell, 2010). In an effort to develop an independent age determination method, X-ray spectroscopy, and laser ablation inductively coupled plasma mass spectrometry (LA-ICP-MS) were used to assess whether seasonal peaks in certain elements were present in these hard structures (Campbell, 2010). While X-ray spectroscopy was unsuccessful, LA-ICP-MS revealed significant correlations between peaks of ^{24}Mg, ^{88}Sr, ^{107}Ag, and ^{109}Ag and banding patterns visible in pectoral girdles and mesoptergia (Campbell, 2010). These methods will require verification and validation to determine whether they can be used for age determination of Bluntnose Sixgill Sharks.

Elemental analyses such as LA-ICP-MS have increasingly been applied to the otoliths of teleost fishes for the purposes of understanding movement patterns, population connectivity, and environmental histories, as well as for validating age estimates (Elsdon et al., 2008; Kalish, 1989; Thorrold et al., 2007). However, there have been relatively few such studies on elasmobranchs. This could be because the processes controlling elemental uptake into the hard parts of elasmobranchs are not as well understood (McMillan et al., 2017). In teleosts, these methods rely on the fact that otoliths are metabolically inert, and that uptake of certain elements is affected by variations in environmental factors such as temperature or salinity (Campana, 2004; Thorrold et al., 2007). In this way, elemental signatures can act as natural tags, reflecting either seasonal changes in ambient conditions or fish movements through water masses with different physical and chemical properties. Raoult et al. (2016) found a correspondence between the number of strontium peaks and the number of annuli counted on sections of vertebral centra for several Australian shark species, though the relationship was not statistically significant. Hale et al. (2006) found a strong positive correlation between the number of calcium peaks and the number of annuli in Round Stingray (*Urobatis halleri*) from southern California, contrary to results of Raoult et al. (2016) for this element. Though elasmobranch vertebrae, like otoliths, are metabolically inert, they are also vascularized (Ferreira and Vooren, 1991), which may physiologically affect the uptake of certain elements. Clearly, more work is needed to understand mechanisms of elemental incorporation into both traditionally used and novel hard structures before these methods are applied as an age validation approach for elasmobranchs.

Another challenge that has been identified in NEP elasmobranch age determination is the potential for banding patterns within hard structures to be a function of fish size rather than of time. For example, in a study by Natanson et al. (2008), Basking Shark (*Cetorhinus maximus*) vertebrae were prepared by X-radiography and gross sectioning. Growth zones were visible, but counts differed depending upon the region of the body from which the vertebral centra were taken (Natanson et al., 2008). The authors concluded that these zones may be related to physiological growth-related factors instead of time. Similar results were found for the Pacific Angel Shark (*Squatina californica*), and were further supported by studying vertebrae of animals that had been injected with OTC (Natanson and Cailliet, 1990). These studies highlight the importance of understanding the periodicity of growth zone formation on a species-specific basis.

Independent methods that do not rely on enumeration of growth zones in hard structures may show greater potential for ageing challenging species. In the North Atlantic, the Greenland Shark (*Somniosus microcephalus*), which is closely related to the Pacific Sleeper Shark, has been aged using bomb radiocarbon analysis applied to eye lenses (Nielsen et al., 2016). Vertebrate eye lenses are metabolically inert, and thus material deposited within the nucleus is representative of conditions the animal was exposed to in early life (Lynnerup et al., 2008; Nielsen et al., 2016). Within the eye lens nuclei of modern-collected sharks, Nielsen et al. (2016) detected elevated ^{14}C levels attributed to the bomb radiocarbon pulse. This increase in ^{14}C was detected only in the smallest immature Greenland Sharks in the Nielsen et al. (2016) study, suggesting an age at maturity of at least 48 years and potential extreme longevity. Bomb radiocarbon analysis could similarly be applied to eye lenses of Pacific Sleeper Sharks, or other species that have been challenging to age using traditional methods.

Another technology that could show promise for age determination of NEP elasmobranchs is near infrared (NIR) spectroscopy (Robins et al., 2015). This nondestructive technique assesses the chemical composition of a material using light absorption from the NIR spectrum (800–2780 nm). The NIR spectra collected from hard structures such as otoliths or vertebrae can then be used to predict fish age. The NIR spectroscopy method requires a calibration based on either known or estimated ages, so it would not be applicable to species for which accurate and precise age determination methods have not been developed. Additional caveats associated with this approach are that it is currently unclear what chemical constituents within hard structures are correlated with the NIR spectra and the mechanism by which they relate to age (Robins et al., 2015). Furthermore, thorough investigations would be required to determine whether NIR would be applicable to older fish as growth slows. However, because the traditional methods for elasmobranch age determination are often labour intensive, NIR spectroscopy is advantageous in that it allows for rapid and cost-effective analysis of large numbers of samples. This method was successfully applied to age-validated vertebrae of Great Hammerhead (*Sphyrna mokarran*) and Spot-tail Shark (*Carcharhinus sorrah*), with strong agreement between the known ages and

NIR spectra ($R^2=0.89$ and 0.84 for the two species, respectively) (Rigby et al., 2016). Because this technology is beginning to become accessible to production age determination laboratories such as at the AFSC, it could mean that larger numbers of fish could be aged to support a greater number of stock assessments if the method is proven to be successful for each species of interest.

4. THE IMPORTANCE OF AGE AND GROWTH DATA TO STOCK ASSESSMENTS AND MANAGEMENT

Fish length and age data are used within stock assessments to describe the growth of a given population. Sometimes, these data are incorporated directly into age-structured population dynamics models (see Kacev et al., in press). In other cases, such as in data-limited situations, they are used to generate age compositions and estimates of life history parameters for informational purposes. Age-structured models divide a stock into individual year classes and follow each year class dynamically through time, incorporating vital rates related to growth, mortality, and recruitment to determine appropriate harvest limits for the population (Quinn and Deriso, 1999). Hence, accurate age data (or known age estimation error rates) on which these year class assignments are based are critical to evaluating population dynamics and estimating yield, as is a general understanding of the underlying growth process of individuals contributing to the stock (Haddon, 2001). The following sections describe how growth models and related life history parameters are derived from age estimates and how age and growth data are currently being used in stock assessments for three common NEP elasmobranch species.

4.1 Growth Models and Associated Life History Parameters: Applications to Stock Assessment

Typically, a suite of growth models is fitted to length-at-age data to better understand the growth of a species. Multiple models are often presented for comparison, although the best fitting and most biologically realistic model can vary by species. Here, we give a brief overview of mathematical models that have been used to describe size-at-age in elasmobranchs, though by no means is it an exhaustive list. See Quinn and Deriso (1999) and Haddon (2001) for more thorough discussions of growth models, model selection, and testing for significant differences among competing growth models. The most commonly used model is the von Bertalanffy three-parameter model (von Bertalanffy, 1938). This model contains a growth coefficient, κ, that describes the average rate at which organisms within a population achieve maximum length from length at birth, and is often referenced as the key parameter to describe a species as having a fast or slow growth rate. The three-parameter von Bertalanffy model often results in the best statistical fit to length-at-age data; however, there is some debate regarding its appropriateness for elasmobranchs because the t_0 parameter is biologically meaningless (Cailliet and Goldman, 2004; Fabens, 1965, and others). A two-parameter formulation that contains a fixed value representing length at birth (L_0) (Fabens, 1965) is often applied in elasmobranch age and growth studies (Cailliet et al., 2006). This parameterization is most appropriate when size at birth is known or sample sizes are small (Goldman et al., 2012).

Other commonly used growth models are the Gompertz model (Ricker, 1975) and the logistic growth curve (Ricker et al., 1979), both of which are *S*-shaped. It has been suggested that sigmoidal functions may be more appropriate for modelling the growth of batoids or other species that grow in volume at a faster rate than in length (Ainsley et al., 2011; Goldman et al., 2012). Other, less common, growth models that may be useful for elasmobranchs include the Richards (1959) model, Schnute (1981) model, and various mathematical fits (e.g., linear, exponential, or polynomial). The Schnute (1981) growth model is essentially a generalized form of the von Bertalanffy, Gompertz, Richards, and logistic models (Quinn and Deriso, 1999). Models that incorporate separate growth phases (e.g., before and after maturity) have also been found to fit length-at-age data well (Araya and Cubillos, 2006; Braccini et al., 2007; Tribuzio et al., 2010).

Selection of the best growth model is generally based on mathematical goodness-of-fit criteria, such as mean-squared error or the Akaike information criterion (Burnham and Anderson, 2004). However, it is also important to use biological reasoning and to consider model assumptions when selecting a model to best describe growth. Chen et al. (1992) examined multiple-order polynomial functions with teleost fish length-at-age data and found that while in some instances they fit the data well, they often produced results that were biologically meaningless (such as negative growth). Similarly, in some cases, two-phase models suggest a slight decline in size-at-age at the transition between phases (Braccini et al., 2007; Tribuzio et al., 2010). If model parameters such as L_∞ are vastly different from maximum observed length, the model may be rejected for lack of biological realism even if it statistically provides the best fit to the length-at-age data.

Growth models are incorporated into stock assessments primarily in two ways: (1) length-at-age and/or weight-at-age transitions (e.g., age-length keys) that provide age and size structure for a population and (2) estimated life history parameters. Life history parameters can include (but are not limited to) natural mortality (M), population growth rates (e.g., the intrinsic rate of population increase, λ or r), and age at maturity. Natural mortality can be estimated by a number of indirect methods, but most commonly rely on the von Bertalanffy growth coefficient, κ (Alverson and Carney, 1975; Jensen, 1996) or maximum age (e.g., Hoenig, 1983). Tribuzio and Kruse (2012) show a variety of methods for estimating M and include discussion of how those methods were derived and may apply to elasmobranchs. The maximum observed age, while commonly reported, may not accurately represent the longevity of the species due to small sample sizes of older specimens; further, it tends to underestimate longevity in any fished population, especially as fishing mortality increases. Sampling constraints and the logistical difficulty associated with ageing large numbers of fish may necessitate calculating the maximum theoretical age (e.g., Taylor, 1958) to estimate M. Natural mortality can be time or age varying, but typically a single point estimate of M is used, and there is often a large amount of uncertainty associated with this parameter. Because growth-related parameters are so closely related to M (Gislason et al., 2010), it is imperative to obtain accurate estimates of these life history traits. Natural mortality is an extremely important life history parameter to stock assessments, as it is combined with estimated fishing mortality to provide the total instantaneous mortality experienced by a stock or population. It also is an indicator of how susceptible the population or stock may be to overfishing. In general, longer-lived, slower-growing species have lower population growth rates and are less capable of recovering from fishing pressure.

A comprehensive list of growth studies and resultant age-related parameters for elasmobranchs in the NEP is provided in Table 1. Where available, von Bertalanffy model parameters are provided to aid in comparison among species. This table contains all parameters that have been reported for the NEP at the time of this writing. As age and growth technology evolves, these parameter estimates may be refined in the future.

4.2 Examples of Age Data in NEP Stock Assessments

Currently, there are relatively few species in the NEP for which age and growth data are used in stock assessments, due to the fact that the amount of age data available for most NEP species is limited to relatively short periods (1–2 years), or is nonexistent. However, in recent years dedicated efforts have been made to regularly or semiregularly age certain common species to support data-based stock assessments. Here, we present three examples of species that are managed using age and growth data: Alaska Skate, Longnose Skate, and North Pacific Spiny Dogfish. Not discussed here are the GOA and BSAI Salmon Shark stock assessments, which report maximum age, age at first recruitment, and von Bertalanffy growth parameters but do not use these values to determine harvest recommendations (Tribuzio et al., 2015, 2016a). Another species not included here is the Blue Shark, assessed by the Highly Migratory Species group of the Pacific Fishery Management Council, which uses an age-structured model for the entire NEP stock (Kleiber et al., 2009); however, the Blue Shark assessment is described in more detail in Kacev et al. (in press).

4.2.1 Alaska Skate

The Alaska Skate is by far the most abundant species of skate in the eastern Bering Sea and is also one of the most abundant groundfish in the region (Lauth, 2010). This species is widely distributed over the continental shelf and is most commonly found at depths ranging from 50 to 200 m (Stevenson, 2004). The Alaska Skate uses several discrete nursery areas that occur along the interface of the continental shelf and slope (Hoff, 2008, 2010), where it deposits pairs of leathery egg cases each containing one embryo. The period from deposition to hatching is estimated to last over 3 years (Hoff, 2008). After hatching, juveniles move from the shelf–slope interface to shallower waters of the Bering Sea, returning to the nursery areas to reproduce after they reach maturity (Hoff, 2008).

The Alaska Skate has been aged from caudal thorns and vertebral centrum sections (Matta and Gunderson, 2007), although centrum sections are the currently preferred method due to higher precision of age estimates. The maximum observed age to date for this species is 26 years (Table 1, M.E. Matta, AFSC, NOAA, unpublished data), older than the previous maximum age of 17 years published in Matta and Gunderson (2007), although the 26-year-old individual was unusually large (137 cm total length, measured from the tip of the snout to the tip of the tail in a straight line). Genetic testing confirmed that at-sea species identification was correct (I. Spies, AFSC, NOAA, personal communication, 2013), making this the largest and oldest positively identified Alaska Skate to date. Male and female Alaska Skate reach 50% maturity at the ages of 9 and 10 years, respectively (Matta and Gunderson, 2007). Age estimates for this species have been verified by edge and marginal increment analyses (Matta and Gunderson, 2007).

Two different growth models have been fitted to length-at-age data for the Alaska Skate: the three-parameter von Bertalanffy model and the Gompertz model (Matta and Gunderson, 2007). Both models were found to fit the data well; however, the Gompertz model had slightly better goodness-of-fit and provided more biologically realistic estimates of asymptotic length (Matta and Gunderson, 2007). Significant differences were found between the sexes in overall length-at-age, although these differences were very slight (Matta and Gunderson, 2007).

The Alaska Skate is managed in the BSAI using an age-structured model that directly incorporates length-at-age (Ormseth, 2016). A four-parameter Schnute growth function (Schnute, 1981) is used to approximate the Gompertz growth function, and growth parameters are estimated directly within the stock assessment model (Ormseth, 2016). Age determination error is also incorporated within the model to account for variability among age readers. The estimate of natural mortality used in the stock assessment model is derived from age-related life history parameters. The model produces numbers of fish at age, which are used in part to generate harvest recommendations. Since the Matta and Gunderson (2007) study, Alaska Skate age samples have been collected on an intermittent basis to continue to update the parameter estimates used in the stock assessment. Samples were collected in 2003, 2007, 2008, 2009, and 2015, and will continue to be collected every 5 years for the foreseeable future. The Alaska Skate is the only elasmobranch species in NEP waters in which an age-structured model, supported by regular age determination, is currently used for stock assessment and harvest recommendations (Ormseth, 2016).

Fish stock status is defined by the NOAA Fisheries Office of Sustainable Fisheries as follows: 'Overfishing' means that the annual catch rate is too high and 'Overfished' means that the population size is too small. These definitions are based on reference points from stock assessments and are used to determine whether current fisheries management practices are adequate. Age data are required to assess whether a stock is in an overfished status; however, age data are not necessary for declaring whether overfishing is occurring. Based on these definitions, the Alaska Skate stock is not currently considered overfished, subject to overfishing, or approaching an overfished state (Ormseth, 2016).

4.2.2 Longnose Skate

The Longnose Skate is a common, large-bodied species (145 cm maximum TL) that ranges from the eastern Bering Sea to Baja California at depths of

9–1069 m (Ebert et al., 2008; Stevenson et al., 2007). While the Longnose Skate is well distributed from the GOA south along the WC, it can occur in large localized aggregations and is most strongly associated with mixed substrates (Bizzarro et al., 2014; Stevenson et al., 2007). Relatively little is known about Longnose Skate biology, although recent work has focused on providing regional estimates of life history parameters. Estimates of length at 50% maturity vary by region, ranging from 72 to 103 cm for males and 90 to 113 cm for females (Ebert et al., 2008; McFarlane and King, 2006). Reproduction is thought to occur year-round (Ebert et al., 2008). Fishes are the most important component of the diet, followed by shrimp and cephalopods (Bizzarro et al., 2007; Robinson et al., 2007).

Longnose Skates have been aged primarily from unstained vertebral centrum sections, although histological centrum sections have also been used (Gburski et al., 2007; McFarlane and King, 2006; Thompson, 2005; Zeiner and Wolf, 1993). Maximum reported ages are 22 years for the WC (Thompson, 2005), 26 years for BC (McFarlane and King, 2006), and 25 years for the GOA (Gburski et al., 2007); however, discrepancies may exist among laboratories in age determination methodology. Based on recent cooperative efforts between the AFSC, the Canadian Department of Fisheries and Oceans (DFO), and the PSRC, Longnose Skate age estimation at AFSC (Gburski et al., 2007) was found to be accurate based on bomb radiocarbon validation (King et al., 2017).

Assessments have been conducted for Longnose Skate in the four main NEP management regions (WC, BC, GOA, and BSAI) with varied usage of age and growth information. The WC assessment is based on an age-structured model (Gertseva and Schirripa, 2008), relying on a number of studies to inform the von Bertalanffy equation used within the model, and does not incorporate new age estimates (i.e., production age determination). Age at maturity and maximum age are also used to generate estimates of natural mortality employed in the model. In contrast, the BC assessment is considered data-limited, and age and growth information is used only to provide life history information (e.g., natural mortality, King et al., 2015). The GOA and BSAI assessments only report age-related values (e.g., maximum observed age), but do not use existing age data to estimate natural mortality, the only parameter relevant to that data-limited assessment (Ormseth, 2015, 2016). The GOA and BSAI stocks are not currently subject to overfishing (Ormseth, 2015, 2016). The WC Longnose Skate stock was not considered overfished as of the last assessment

(Gertseva and Schirripa, 2008), and insufficient information exists to determine the stock status of BC Longnose Skate (King et al., 2015).

4.2.3 North Pacific Spiny Dogfish

North Pacific Spiny Dogfish occupy continental shelf and upper slope waters from the Bering Sea to the Baja Peninsula in the eastern North Pacific and south through the Japanese archipelago in the western North Pacific. They are considered more common off the WC and BC than in the GOA or BSAI (Hart, 1973; Ketchen, 1986; Mecklenburg et al., 2002). In Alaska, they are more common in the GOA than in the BSAI. North Pacific Spiny Dogfish inhabit both benthic and pelagic environments to a maximum recorded depth of 677 m (C.A. Tribuzio, AFSC, NOAA, unpublished data). North Pacific Spiny Dogfish are commonly found in the water column and in surface waters (C.A. Tribuzio, AFSC, NOAA, unpublished data).

North Pacific Spiny Dogfish in the NEP grow to a maximum total length of 160 cm (Compagno, 1984). A recent study in the GOA estimated ages-at-50% maturity to be 36 years for females and 21 years for males (Tribuzio and Kruse, 2012), which is similar to estimates from BC of 35 and 19 years, respectively (Saunders and McFarlane, 1993). Longevity in the NEP is between 80 and 100 years (Campana et al., 2006). Growth coefficients (κ) for this species are among the lowest of all shark species, $\kappa = 0.03$ for females and 0.06 for males (Tribuzio et al., 2010). Age estimates for this species have been validated through bomb radiocarbon analysis (Campana et al., 2006) and mark–recapture analysis (Beamish and McFarlane, 1985; McFarlane and Beamish, 1987; McFarlane and King, 2009).

Similar to Longnose Skate, North Pacific Spiny Dogfish assessments are conducted for the WC, BC, GOA, and BSAI. However, age data are only used to inform harvest recommendations in the WC and GOA assessments. The WC assessment uses an age-structured model, though the stock assessment scientists acknowledge the uncertainties associated with the age data (Taylor et al., 2013). To address this, age and growth parameters are used to inform the model (i.e., set initial parameter values), but are not included as part of the model input data (Gertseva and Taylor, 2012). The GOA assessment is considered data limited, and age data are used to derive life history parameters, such as natural mortality, to determine harvest recommendations (Tribuzio et al., 2015). These stocks are not currently considered subject to overfishing (Gertseva and Taylor, 2012; Tribuzio et al., 2015, 2016a).

5. CONCLUSIONS

The purpose of this chapter was to profile how age and growth data are currently collected and used for conservation and management of NEP elasmobranch species, as well as to provide information on advancements in this field. Across the globe, the field of age determination continues to evolve rapidly, both for teleost and cartilaginous fishes; thus we recommend that researchers expand their literature searches to include species that are not necessarily endemic to the NEP when investigating new potential ageing methods. Most critical is the need to validate age estimates, because if estimates are inaccurate, they have the potential to propagate serious errors throughout stock assessment models, which could in turn lead to overfishing.

While the number of studied species has dramatically increased in recent years, more work is needed to assess the accuracy of the various methods profiled here and to develop new technologies in cases where traditional methods are not successful. Given the life history characteristics of elasmobranchs, their relative vulnerability to fishing mortality, and the potential for increased interest in the development of shark and batoid fisheries, we expect that new stock assessments will be conducted for NEP elasmobranch species as biological and catch information improves. These assessments will be reliant on estimates of age and related life history parameters.

Ultimately, the purpose of conducting age and growth studies is to gain insight into the biology of the species and inform stock assessments. Most current elasmobranch stock assessments are considered data limited, relying heavily on catch history or survey trends that may not sample the population well. Thus, improving the knowledge of age and growth is crucial to advancing management of fisheries that interact with elasmobranchs, and to promote conservation of the species.

ACKNOWLEDGEMENTS

The authors of this review would like to thank all of the scientists and students who have contributed to bettering our understanding of elasmobranch life histories. In particular, we thank Greg Cailliet and the numerous former and current graduate students at the Pacific Shark Research Center (Moss Landing, CA) who have pioneered much of this research. We thank Gordon McFarlane and Jackie King (Department of Fish and Oceans, Canada), Duane Stevenson (AFSC), and Steve Campana (University of Iceland) for allowing us to reproduce their images. Finally, we thank Steve Campana, Chris Harvey-Clark (Dalhousie

University), Shawn Larson (Seattle Aquarium), and Dayv Lowry (Washington Department of Fish and Wildlife) for providing helpful reviews that greatly improved this chapter. Reference to trade names does not imply endorsement by the National Marine Fisheries Service, NOAA. The findings and conclusions in the paper are those of the authors and do not necessarily represent the views of the National Marine Fisheries Service, NOAA.

REFERENCES

Ainsley, S.M., Ebert, D.A., Cailliet, G.M., 2011. Age, growth, and maturity of the Whitebrow Skate, *Bathyraja minispinosa*, from the eastern Bering Sea. ICES J. Mar. Sci. 68, 1426–1434.

Ainsley, S.M., Ebert, D.A., Natanson, L.J., Cailliet, G.M., 2014. A comparison of age and growth of the Bering Skate, *Bathyraja interrupta* (Gill and Townsend, 1897), from two Alaskan large marine ecosystems. Fish. Res. 154, 17–25.

Alverson, D.L., Carney, M.J., 1975. A graphic review of the growth and decay of population cohorts. ICES J. Mar. Sci. 36, 133–143.

Andrews, A.H., Kerr, L.A., 2015. Validated age estimates for large white sharks of the northeastern Pacific Ocean: altered perceptions of vertebral growth shed light on complicated bomb $\Delta^{14}C$ results. Environ. Biol. Fishes 98, 971–978.

Araya, M., Cubillos, L.A., 2006. Evidence of two-phase growth in elasmobranchs. Environ. Biol. Fishes 77, 293–300.

Barnett, L.A., Ebert, D.A., Cailliet, G.M., 2009. Assessment of the dorsal fin spine for chimaeroid (Holocephali: Chimaeriformes) age estimation. J. Fish Biol. 75, 1258–1270.

Beamish, R.J., McFarlane, G.A., 1985. Annulus development on the second dorsal spine of the Spiny Dogfish (*Squalus acanthias*) and its validity for age determination. Can. J. Fish. Aquat. Sci. 42, 1799–1805.

Bizzarro, J.J., Robinson, H.J., Rinewalt, C.S., Ebert, D.A., 2007. Comparative feeding ecology of four sympatric skate species off central California, USA, Environ. Biol. Fish. 80, 197–220.

Bizzarro, J.J., Broms, K.M., Logsdon, M.G., Ebert, D.A., Yoklavich, M.M., Kuhnz, L.A., Summers, A.P., 2014. Spatial segregation in eastern North Pacific skate assemblages. PLoS One 9, e109907.

Braccini, J.M., Gillanders, B.M., Walker, T.I., 2007. Comparison of deterministic growth models fitted to length-at-age of the Piked Spurdog (*Squalus megalops*) in south-eastern Australia. Mar. Freshw. Res. 58, 24–33.

Braccini, J.M., Troynikov, V.S., Walker, T.I., Mollet, H.F., Ebert, D.A., Barnett, A., Kirby, N., 2010. Incorporating heterogeneity into growth analyses of wild and captive broadnose Sevengill Sharks *Notorynchus cepedianus*. Aquatic Biol. 9, 131–138.

Bubley, W.J., Kneebone, J., Sulikowski, J.A., Tsang, P.C.W., 2012. Reassessment of Spiny Dogfish *Squalus acanthias* age and growth using vertebrae and dorsal-fin spines. J. Fish Biol. 80, 1300–1319.

Burnham, K.P., Anderson, D.R., 2004. Multimodel inference: understanding AIC and BIC in model selection. Sociol. Method. Res. 33, 261–304.

Cailliet, G., 1990. In: Pratt, H.L., Gruber, S.H., Taniuchi, T. (Eds.), Elasmobranch age determination and verification: an updated review. Elasmobranchs as Living Resources: Advances in the Biology, Ecology, Systematics, and the Status of the Fisheries. NOAA Tech. Rep, p. 90.

Cailliet, G.M., 2015. Perspectives on elasmobranch life-history studies: a focus on age validation and relevance to fishery management. J. Fish Biol. 87, 1271–1292.

Cailliet, G.M., Goldman, K.J., 2004. Age determination and validation in chondrichthyan fishes. In: Carrier, J.C., Musick, J.A., Heithaus, M.R. (Eds.), Biology of Sharks and Their Relatives. CRC Press, Boca Raton, FL, pp. 399–447.

Cailliet, G., Martin, L.K., Harvey, J.T., Kusher, D., Welden, B.A., 1983. In: Prince, E.D., Pulos, L.M. (Eds.), Preliminary studies on the age and growth of Blue, *Prionace glauca*, Common Thresher, *Alopias vulpinus* and Shortfin Mako, *Isurus oxyrinchus*, Sharks from California waters. Proceedings of the International Workshop on Age Determination of Oceanic Pelagic Fishes: Tunas, Billfishes, and Sharks. NOAA Tech. Rep, p. 8.

Cailliet, G.M., Smith, W.D., Mollet, H.F., Goldman, K.J., 2006. Age and growth studies of chondrichthyan fishes: the need for consistency in terminology, verification, validation, and growth function fitting. Environ. Biol. Fishes 77, 211–228.

Camhi, M., Fowler, S., Musick, J., Brautigam, A., Fordham, S., 1998. Sharks and their relatives: ecology and conservation. Occasional Paper No. 20, 39 pp.

Campana, S.E., 2001. Accuracy, precision and quality control in age determination, including a review of the use and abuse of age validation methods. J. Fish Biol. 59, 197–242.

Campana, S.E., 2004. Otolith elemental composition as a natural marker of fish stocks. In: Cadrin, S.X., Friedland, K.D., Waldman, J.R. (Eds.), Stock Identification Methods: Applications in Fishery Science. Elsevier Academic Press, Amsterdam, pp. 227–245.

Campana, S.E., 2014. Age determination of elasmobranchs, with special reference to Mediterranean species: a technical manual. General Fisheries Commission for the Mediterranean, FAO No. 94, 38 pp.

Campana, S.E., Jones, C., McFarlane, G.A., Myklevoll, S., 2006. Bomb dating and age validation using the spines of Spiny Dogfish (*Squalus acanthias*). Environ. Biol. Fishes 77, 327–336.

Campbell, J.S., 2010. Age determination of the Sixgill Shark from hard parts using a series of traditional and novel approaches. M.S. Thesis, Western Washington University, Bellingham, WA.

Chen, Y., Jackson, D.A., Harvey, H.H., 1992. A comparison of the von Bertalanffy and polynomial functions in modelling fish growth data. Can. J. Fish. Aquat. Sci. 49, 1228–1235.

Compagno, L.J.V., 1984. Sharks of the world: an annotated and illustrated catalogue of shark species known to date. FAO Fish. Synop. No. 125, Vol. 4, 675 pp.

Conrath, C.L., Gelsleichter, J.B., Musick, J.A., 2002. Age and growth of the Smooth Dogfish (*Mustelus canis*) in the northwest Atlantic Ocean. Fish. Bull. 100, 674–682.

Cortes, E., 2002. Incorporating uncertainty into demographic modeling: application to shark populations and their conservation. Conserv. Biol. 16, 1048–1062.

Davis, C.D., Cailliet, G.M., Ebert, D.A., 2007. Age and growth of the Roughtail Skate *Bathyraja trachura* (Gilbert 1892) from the eastern North Pacific. Environ. Biol. Fishes 80, 325–336.

Dean, M.N., Summers, A.P., 2006. Mineralized cartilage in the skeleton of chondrichthyan fishes. Fortschr. Zool. 109, 164–168.

Ebert, D.A., Smith, W.D., Cailliet, G.M., 2008. Reproductive biology of two commercially exploited skates, *Raja binoculata* and *R. rhina*, in the western Gulf of Alaska. Fish. Res. 94, 48–57.

Elsdon, T.S., Wells, B.K., Campana, S.E., Gillanders, B.M., Jones, C.M., Limburg, K.E., Secor, D.H., Thorrold, S.R., Walther, B.D., 2008. Otolith chemistry to describe movements and life-history parameters of fishes: hypotheses, assumptions, limitations and inferences. Oceanogr. Mar. Biol. 46, 297–330.

Fabens, A.J., 1965. Properties and fitting of the von Bertalanffy growth curve. Growth 29, 265–289.

Ferreira, B.P., Vooren, C.M., 1991. Age, growth, and structure of vertebra in the School Shark *Galeorhinus galeus* (Linnaeus, 1758) from southern Brazil. Fish. Bull. 89, 19–31.

Francis, M.P., Campana, S.E., Jones, C.M., 2007. Age under-estimation in New Zealand Porbeagle Sharks (*Lamna nasus*): is there an upper limit to ages that can be determined from shark vertebrae? Mar. Freshw. Res. 58, 10–23.

Gallagher, M., Nolan, C.P., 1999. A novel method for the estimation of age and growth in rajids using caudal thorns. Can. J. Fish. Aquat. Sci. 56, 1590–1599.

Gallagher, M.J., Green, M.J., Nolan, C.P., 2006. The potential use of caudal thorns as a non-invasive ageing structure in the Thorny Skate (*Amblyraja radiata* Donovan, 1808). Environ. Biol. Fishes 77, 265–272.

Gburski, C.M., Gaichas, S.K., Kimura, D.K., 2007. Age and growth of Big Skate (*Raja binoculata*) and Longnose Skate (*R. rhina*) in the Gulf of Alaska. Environ. Biol. Fishes 80, 337–349.

Gertseva, V., Schirripa, M.J., 2008. Status of the Longnose Skate (*Raja rhina*) off the continental US Pacific Coast in 2007. 131 pp.

Gertseva, V.V., Taylor, I.G., 2012. Status of the Spiny Dogfish shark resource off the continental U.S. Pacific Coast in 2011. 259 pp.

Gislason, H., Daan, N., Rice, J.C., Pope, J.G., 2010. Size, growth, temperature and the natural mortality of marine fish. Fish Fish. 11, 149–158.

Goldman, K.J., 2005. Age and growth of elasmobranch fishes. In: Musick, J.A., Bonfil, R. (Eds.), Elasmobranch Fisheries Management Techniques. IUCN Shark Specialist Group and Asia Pacific Economic Cooperation (APEC), Rome, Italy, pp. 97–132.

Goldman, K.J., Musick, J.A., 2006. Growth and maturity of Salmon Sharks (*Lamna ditropis*) in the eastern and western North Pacific, and comments on back-calculation methods. Fish. Bull. 104, 278–292.

Goldman, K.J., Cailliet, G., Andrews, A.H., Natanson, L.J., 2012. Assessing the age and growth of chondrichthyan fishes. In: Carrier, J.C., Musick, J.A., Heithaus, M.R. (Eds.), Biology of Sharks and Their Relatives. CRC Press, Boca Raton, FL, pp. 423–451.

Haas, D.L., Ebert, D.A., Cailliet, G.M., 2016. Comparative age and growth of the Aleutian Skate, *Bathyraja aleutica*, from the eastern Bering Sea and Gulf of Alaska. Environ. Biol. Fishes 99, 813–828.

Haddon, M., 2001. Modelling and Quantitative Methods in Fisheries. Chapman & Hall/ CRC Press, Boca Raton, FL.

Hale, L.F., Dudgeon, J.V., Mason, A.Z., Lowe, C.G., 2006. Elemental signatures in the vertebral cartilage of the Round Stingray, *Urobatis halleri*, from Seal Beach. Calif. Environ. Biol. Fish. 77, 317–325.

Hamady, L.L., Natanson, L.J., Skomal, G.B., Thorrold, S.R., 2014. Vertebral bomb radiocarbon suggests extreme longevity in White Sharks. PLoS One 9, e84006.

Hart, J.L., 1973. Pacific fishes of Canada. Fish. Res. Bd. Can. Bull., 180. Ottawa, Canada.

Henderson, A.C., Arkhipkin, A.I., Chtcherbich, J.N., 2004. Distribution, growth and reproduction of the White-spotted Skate *Bathyraja albomaculata* (Norman, 1937) around the Falkland Islands. J. Northw. Atl. Fish. Sci. 35, 79–87.

Hoenig, J.M., 1983. Empirical use of longevity data to estimate mortality rates. Fish. Bull. 82, 898–903.

Hoenig, J.M., Brown, C.A., 1988. A simple technique for staining growth bands in elasmobranch vertebrae. Bull. Mar. Sci. 42, 334–337.

Hoenig, J.M., Gruber, S.H., 1990. In: Pratt, H.L., Gruber, S.H., Taniuchi, T. (Eds.), Life-history patterns in the elasmobranchs: implications for fisheries managment. Elasmobranchs as Living Resources: Advances in the Biology, Ecology, Systematics, and the Status of Fisheries. NOAA Tech. Rep, p. 90.

Hoff, G.R., 2008. A nursery site of the Alaska Skate (*Bathyraja parmifera*) in the eastern Bering Sea. Fish. Bull. 106, 233–244.

Hoff, G.R., 2010. Identification of skate nursery habitat in the eastern Bering Sea. Mar. Ecol. Prog. Ser. 403, 243–254.

James, K.C., Ebert, D.A., Natanson, L.J., Cailliet, G.M., 2014. Age and growth characteristics of the Starry Skate, *Raja stellulata*, with a description of life history and habitat trends of the central California, U.S.A., skate assemblage. Environ. Biol. Fishes 97, 435–448.

Jensen, A.L., 1996. Beverton and Holt life history invariants result from optimal trade-off of reproduction and survival. Can. J. Fish. Aquat. Sci. 53, 820–822.

Kacev D., Sippel T.J., Kinney M.J., Pardo S.A., Mull C.G. An introduction to modeling abundance and life-history parameters in shark populations. Northeast Pacific Shark Biology, Research and Conservation Part B, Advances in Marine Biology, vol. 78. In press.

Kalish, J.M., 1989. Otolith microchemistry: validation of the effects of physiology, age and environment on otolith composition. J. Exp. Mar. Biol. Ecol. 132, 151–178.

Kalish, J.M., 1993. Pre- and post-bomb radiocarbon in fish otoliths. Earth Planet. Sci. Lett. 114, 549–554.

Kerr, L.A., Andrews, A.H., Frantz, B.R., Coale, K.H., Brown, T.A., Cailliet, G.M., 2004. Radiocarbon in otoliths of Yelloweye Rockfish (*Sebastes ruberrimus*): a reference time series for the coastal waters of southeast Alaska. Can. J. Fish. Aquat. Sci. 61, 443–451.

Ketchen, K.S., 1975. Age and growth of dogfish *Squalus acanthias* in British Columbia waters. J. Fish. Res. Bd. Can. 32, 43–59.

Ketchen, K.S., 1986. The Spiny Dogfish (*Squalus acanthias*) in the Northeast Pacific and a History of its Utilization. Can. Spec. Pub. Fish. Aquat. Sci. 0706-6481, 78 pp.

Kimura, D.K., Kastelle, C.R., Goetz, B.J., Gburski, C.M., Buslov, A.V., 2006. Corroborating the ages of Walleye Pollock (*Theragra chalcogramma*). Mar. Freshw. Res. 57, 323–332.

Kimura, D.K., Anderl, D.M., Goetz, B.J., 2007. Seasonal marginal growth on otoliths of seven Alaska groundfish species support the existence of annual patterns. Alaska Fish. Res. Bull. 12, 243–251.

King, J.R., McFarlane, G.A., 2003. Marine fish life history strategies: applications to fishery management. Fish. Manag. Ecol. 10, 249–264.

King, J.R., McPhie, R.P., 2015. Preliminary age, growth and maturity estimates of Spotted Ratfish (*Hydrolagus colliei*) in British Columbia. Deep-Sea Res. 115 (Pt. II), 55–63.

King, J.R., Surry, A.M., Garcia, S., Starr, P.J., 2015. Big Skate (*Raja binoculata*) and Longnose Skate (*R. rhina*) Stock Assessments for British Columbia. DFO Can. Sci. Advis. Sec. Res. Doc. 2 2015/070, 329.

King, J.R., Helser, T., Gburski, C.M., Ebert, D.A., Cailliet, G.M., Kastelle, C.R., 2017. Bomb radiocarbon analyses validate and inform age determination of Longnose Skate (*Raja rhina*) and Big Skate (*Beringraja binoculata*) in the North Pacific Ocean. Fish. Res. 193, 195–206.

Kleiber, P., Clarke, S., Bigelow, K., Nakano, H., McAllister, M., Takeuchi, Y., 2009. North Pacfic Blue Shark Stock Assessment. U.S. Dep. Commer., NOAA Tech. Memo. NOAA-TM-NMFS-PIFSC-17, 74 pp.

Kusher, D.I., Smith, S.E., Cailliet, G.M., 1992. Validated age and growth of the Leopard Shark, *Triakis semifasciata*, with comments on reproduction. Environ. Biol. Fishes 35, 187–203.

Lauth, R.R., 2010. Results of the 2009 Eastern Bering Sea Continental Shelf Bottom Trawl Survey of Groundfish and Invertebrate Resources. U.S. Dep. Commer., NOAA Tech. Memo. NMFS-AFSC-204, 228 pp.

Lynnerup, N., Kjeldsen, H., Heegaard, S., Jacobsen, C., Heinemeier, J., 2008. Radiocarbon dating of the human eye lens crystallines reveal proteins without carbon turnover throughout life. PLoS One 3, e1529.

Matta, M.E., Gunderson, D.R., 2007. Age, growth, maturity, and mortality of the Alaska Skate, *Bathyraja parmifera*, in the eastern Bering Sea. Environ. Biol. Fishes 80, 309–323.

Maurer, J.R.F., 2009. Life history of two Bering Sea Slope Skates: *Bathyraja lindbergi* and *B. maculata*. M.S. Thesis, California State University, Monterey Bay, CA.

McFarlane, G.A., Beamish, R.J., 1987. Validation of the dorsal spine method of age determination for spiny dogfish. In: Summerfelt, R.C., Hall, G.E. (Eds.), Age and Growth of Fish. Iowa State University Press, Ames, Iowa, pp. 287–300.

McFarlane, G.A., King, J.R., 2006. Age and growth of Big Skate (*Raja binoculata*) and longnose skate (*Raja rhina*) in British Columbia waters. Fish. Res. 78, 169–178.

McFarlane, G.A., King, J.R., 2009. Re-evaluating the age determination of spiny dogfish (*Squalus acanthias*) using oxytetracycline and fish at liberty up to twenty years. In: Gallucci, V.F., McFarlane, G.A., Bargmann, G.G. (Eds.), Biology and Management of Dogfish Sharks. American Fisheries Society, Bethesda, MD, pp. 77–88.

McFarlane, G.A., King, J.R., Saunders, M.W., 2002. Preliminary study on the use of neural arches in the age determination of Bluntnose Sixgill Sharks (*Hexanchus griseus*). Fish. Bull. 100, 861–864.

McMillan, M.N., Izzo, C., Wade, B., Gillanders, B.M., 2017. Elements and elasmobranchs: hypotheses, assumptions and limitations of elemental analysis. J. Fish Biol. 90, 559–594.

McPhie, R.P., Campana, S.E., 2009. Bomb dating and age determination of skates (family Rajidae) off the eastern coast of Canada. ICES J. Mar. Sci. 66, 546–560.

Mecklenburg, C.W., Anthony, T.A., Thorsteinson, L.K., 2002. Fishes of Alaska. American Fisheries Society, Bethesda, MD.

Nakano, H., 1994. Age, reproduction and migration of Blue Shark in the North Pacific Ocean. Bull. Nat. Res. Inst. Far Seas Fish. 31, 141–256.

Natanson, L.J., Cailliet, G.M., 1990. Vertebral growth zone deposition in Pacific Angel Sharks. Copeia 1990, 1133–1145.

Natanson, L.J., Casey, J.G., Kohler, N.E., 1995. Age and growth estimates for the Dusky Shark, *Carcharhinus obscurus*, in the western North Atlantic Ocean. Fish. Bull. 93, 116–126.

Natanson, L.J., Sulikowski, J.A., Kneebone, J.R., Tsang, P.C., 2007. Age and growth estimates for the Smooth Skate, *Malacoraja senta*, in the Gulf of Maine. Environ. Biol. Fishes 80, 293–308.

Natanson, L.J., Wintner, S.P., Johansson, F., Piercy, A., Campbell, P., De Maddalena, A., Gulak, S.J.B., Human, B., Fulgosi, F.C., Ebert, D.A., Hemida, F., Mollen, F.H., Vanni, S., Burgess, G.H., Compagno, L.J.V., Wedderburn-Maxwell, A., 2008. Ontogenetic vertebral growth patterns in the Basking Shark *Cetorhinus maximus*. Mar. Ecol. Prog. Ser. 361, 267–278.

Natanson, L.J., Hamady, L.L., Gervelis, B.J., 2016. Analysis of bomb radiocarbon data for common thresher sharks, *Alopias vulpinus*, in the northwestern Atlantic Ocean with revised growth curves. Environ. Biol. Fishes 99, 39–47.

Neer, J.A., Cailliet, G.M., 2001. Aspects of the life history of the Pacific Electric Ray, *Torpedo californica* (Ayres). Copeia 2001, 842–847.

Nielsen, J., Hedeholm, R.B., Heinemeirer, J., Bushnell, P.G., Christiansen, J.S., Olsen, J., Ramsey, C.B., Brill, R.W., Simon, M., Steffensen, K.F., Steffensen, J.F., 2016. Eye lens radiocarbon reveals centuries of longevity in the greenland shark (*Somniosus microcephalus*). Science 353, 702–704.

Ormseth, 2015. Assessment of the Skate Stock Complex in the Gulf of Alaska, Stock Assessment and Fishery Evaluation Reports for the Groundfish Resources of the Gulf of Alaska. North Pacific Fishery Management Council, Anchorage, AK, pp. 1497–1546.

Ormseth, 2016. Assessment of the Skate Stock Complex in the Bering Sea and Aleutian Islands, Stock Assessment and Fishery Evaluation Reports for the Groundfish Resources of the Bering Sea and Aleutian Islands. North Pacific Fishery Management Council, Anchorage, AK, pp. 1769–1868.

Panfili, J., de Pontual, H., Troadec, H., Wright, P.J., 2002. Manual of Fish Sclerochronology. Ifremer-IRD, Brest, France.

Parsons, G.R., 1983. An examination of the vertebral rings of the Atlantic Sharpnose Shark, *Rhizoprionodon terraenovae*. Northw. Gulf Sci. 6, 63–66.

Passerotti, M.S., Andrews, A.H., Carlson, J.K., Wintner, S.P., Goldman, K.J., Natanson, L.J., 2014. Maximum age and missing time in the vertebrae of Sand Tiger Shark (*Carcharias taurus*): validated lifespan from bomb radiocarbon dating in the western North Atlantic and southwestern Indian Oceans. Mar. Freshw. Res. 65, 674–687.

Perez, C.A.R., Cailliet, G.M., Ebert, D.A., 2011. Age and growth of the Sandpaper Skate, *Bathyraja kincaidii*, using vertebral centra, with an investigation of caudal thorns. J. Mar. Biol. Assoc. U.K. 91, 1149–1156.

Piner, K.R., Wischniowski, S.G., 2004. Pacific Halibut chronology of bomb radiocarbon in otoliths from 1944 to 1981 and a validation of ageing methods. J. Fish Biol. 64, 1060–1071.

Provost, C.M., 2016. Age, growth, and sexual maturity of the Deepsea Skate *Bathyraja abyssicola*. M.S. Thesis, Alaska Pacific University, Anchorage, AK.

Quinn, T.J., Deriso, R.B., 1999. Quantitative Fish Dynamics. Oxford University Press, New York.

Raoult, V., Peddemors, V.M., Zahra, D., Howell, N., Howard, D.L., de Jong, M.D., Williamson, J.E., 2016. Strontium mineralization of shark vertebrae. Sci. Rep. 6, 29698.

Richards, F.J., 1959. A flexible growth function for empirical use. J. Exp. Bot. 10, 290–300.

Ricker, W.E., 1975. Computation and interpretation of biological statistics of fish populations. Bull. Fish. Res. Bd. Can. 191, 1–382.

Ricker, W.E., 1979. Growth rates and models. In: Hoar, W.S., Randall, D.J., Brett, J.R. (Eds.), *Fish Physiology*. Academic Press, New York, NY, pp. 677–743.

Rigby, C.L., Wedding, B.B., Grauf, S., Simpfendorfer, C.A., 2016. Novel method for shark age estimation using near infrared spectroscopy. Mar. Freshw. Res. 67, 537–545.

Robins, J.B., Wedding, B.B., Wright, C., Grauf, S., Sellin, M., Fowler, A., Saunders, T., Newman, S., 2015. Revolutionising Fish Ageing: Using near Infrared Spectroscopy to Age Fish. Department of Agriculture, Fisheries and Forestry, Brisbane, Australia.

Robinson, H.J., Cailliet, G., Ebert, D.A., 2007. Food habits of the Longnose Skate, *Raja rhina* (Jordan and Gilbert 1880), in central California waters. Environ. Biol. Fishes 80, 165–179.

Saunders, M.W., McFarlane, G.A., 1993. Age and length at maturity of the female Spiny Dogfish, *Squalus acanthias*, in the Strait of Georgia, British Columbia, Canada. Environ. Biol. Fish. 38, 49–57.

Schnute, J., 1981. A versatile growth model with statistically stable parameters. Can. J. Fish. Aquat. Sci. 38, 1128–1140.

Serra-Pereira, B., Figueiredo, I., Farias, I., Moura, T., Gordo, L.S., 2008. Description of dermal denticles from the caudal region of *Raja clavata* and their use for the estimation of age and growth. ICES J. Mar. Sci. 65, 1701–1709.

Smith, W.D., Cailliet, G.M., Melendez, E.M., 2007. Maturity and growth characteristics of a commercially exploited stingray, *Dasyatis dipterura*. Mar. Freshw. Res. 58, 54–66.

Smith, S.E., Rasmussen, R.C., Ramon, D.A., Cailliet, G.M., 2008. Biology and ecology of thresher sharks (Alopiidae). In: Camhi, M.D., Pikitch, E.K., Babcock, E.A. (Eds.), Sharks of the Open Ocean: Biology, Fisheries and Conservation. Blackwell Publishing, Oxford, England, pp. 60–68.

Stevens, J.D., 1975. Vertebral rings as a means of age determination in the Blue Shark (*Prionace glauca* L.). J. Mar. Biol. Assoc. U.K. 55, 657–665.

Stevenson, D.E., 2004. Identification of Skates, Sculpins, and Smelts by Observers in North Pacific Groundfish Fisheries (2002–2003). U.S. Dep. Commer., NOAA Tech Memo. NMFS-AFSC-142, 67 pp.

Stevenson, D.E., Orr, J.W., Hoff, G.R., McEachran, J.D., 2007. Field Guide to Sharks, Skates, and Ratfish of Alaska. Alaska Sea Grant College Program. University of Alaska Fairbanks, Fairbanks, Alaska.

Sulikowski, J.A., Morin, M.D., Suk, S.H., Howell, W.H., 2003. Age and growth estimates of the Winter Skate (*Leucoraja ocellata*) in the western Gulf of Maine. Fish. Bull. 101, 405–413.

Sulikowski, J.A., Kneebone, J., Elzey, S., Jurek, J., Danley, P.D., Howell, W.H., Tsang, P.C.W., 2005. Age and growth estimates of the Thorny Skate (*Amblyraja radiata*) in the western Gulf of Maine. Fish. Bull. 103, 161–168.

Tanaka, S., Cailliet, G.M., Yudin, K.G., 1990. In: Pratt, H.L., Gruber, S.H., Taniuchi, T. (Eds.), Differences in growth of the Blue Shark, *Prionace glauca*: technique or population? Elasmobranchs as Living Resources: Advances in the Biology, Ecology, Systematics, and the Status of Fisheries. NOAA Tech. Rep, p. 90.

Taylor, C.C., 1958. Cod growth and temperature. ICES J. Mar. Sci. 23, 366–370.

Taylor, I.G., Gertseva, V., Matson, S.E., 2013. Spine-based ageing methods in the Spiny Dogfish Shark, *Squalus suckleyi*: how they measure up. Fish. Res. 147, 83–92.

Thompson, J.E., 2005. Age, growth and maturity of the Longnose Skate (*Raja rhina*) for the U.S. West Coast and sensitivity to fishing impacts. M.S. Thesis, Oregon State University, Corvallis, OR.

Thorrold, S.R., Zacherl, D.C., Levin, L.A., 2007. Population connectivity and larval dispersal using geochemical signatures in calcified structures. Oceanography 20, 80–89.

Tribuzio, C.A., Kruse, G.H., 2012. Life history characteristics of a lightly exploited stock of *Squalus suckleyi*. J. Fish Biol. 80, 1159–1180.

Tribuzio, C.A., Kruse, G.H., Fujioka, J.T., 2010. Age and growth of Spiny Dogfish (*Squalus acanthias*) in the Gulf of Alaska: analysis of alternative growth models. Fish. Bull. 108, 119–135.

Tribuzio, C., Rodgveller, C., Echave, K., Hulson, P., 2015. Assessment of the Shark Stock Complex in the Gulf of Alaska, Stock Assessment and Fishery Evaluation Reports for the Groundfish Resources of the Gulf of Alaska. North Pacific Fishery Management Council, Anchorage, AK, pp. 1569–1642.

Tribuzio, C., Echave, K., Rodgveller, C., Hulson, P., 2016a. Assessment of the Shark Stock Complex in the Bering Sea and Aleutian Islands, Stock Assessment and Fishery Evaluation Reports for the Groundfish Resources of the Bering Sea and Aleutian Islands. North Pacific Fishery Management Council, Anchorage, AK, pp. 1915–1962.

Tribuzio, C.A., Matta, M.E., Gburski, C.M., Atkins, N., Bubley, W.J., 2016b. Methods for the preparation of Spiny Dogfish (*Squalus suckleyi*) fin spines and vertebrae and an overview of age determinations. Mar. Fish. Rev. 78, 1–13.

Tribuzio, C.A., Matta, M.E., Gburski, C.M., Blood, C., Bubley, W.J., Kruse, G.H., in press. Are Pacific Spiny Dogfish lying about their age? A comparison of ageing structures for *Squalus suckleyi*. Mar. Freshw. Res.

von Bertalanffy, L., 1938. A quantitative theory of organic growth (inquiries on growth laws II). Human Biol. 10, 181–213.

Winemiller, K.O., 2005. Life history strategies, population regulation, and implications for fisheries management. Can. J. Fish. Aquat. Sci. 62, 872–885.

Winemiller, K.O., Rose, K.A., 1992. Patterns of life-history diversification in North American fishes: implications for population regulation. Can. J. Fish. Aquat. Sci. 49, 2196–2218.

Winton, M.V., 2011. Age, growth, and demography of the Roughtail Skate, *Bathyraja trachura* (Gilbert, 1892), from the eastern Bering Sea. M.S. Thesis, California State University, Monterey Bay, CA.

Winton, M.V., Natanson, L.J., Kneebone, J., Cailliet, G.M., Ebert, D.A., 2014. Life history of *Bathyraja trachura* from the eastern Bering Sea, with evidence of latitudinal variation in a deep-sea skate species. J. Mar. Biol. Assoc. U.K. 94, 411–422.

Wischniowski, S., 2008. Technique development for age determination of the Pacific Sleeper Shark (*Somniosus pacificus*). IPHC Rep. Assess. Res. Activities 2008, 389–392.

Yudin, K.G., Cailliet, G.M., 1990. Age and growth of the Gray Smoothhound, *Mustelus californicus*, and the Brown Smoothhound, *Mustelus henlei*, sharks from Central California. Copeia 1990, 191–204.

Zeiner, S.J., Wolf, P., 1993. In: Branstetter, S. (Ed.), Growth characteristics and estimates of age at maturity of two species of skates (*Raja binoculata* and *Raja rhina*) from Monterey Bay, California. Conservation Biology of Elasmobranchs. NOAA NMFS Tech. Rep, p. 115.

SUBJECT INDEX

Note: Page numbers followed by "*f*" indicate figures, and "*t*" indicate tables.

O

P

TAXONOMIC INDEX

Note: Page numbers followed by "*f*" indicate figures, and "*t*" indicate tables.

T

U

Z